AN INTRODUCTION TO HELMINTHES

AN INTRODUCTION TO HELMINTHES

By

H.S. Bhamrah
Kavita Juneja

Edited by

Balvinder K.

ANMOL PUBLICATIONS PVT. LTD.
NEW DELHI - 110 002 (INDIA)

ANMOL PUBLICATIONS PVT. LTD.
4374/4B, Ansari Road, Daryaganj
New Delhi - 110 002

An Introduction to Helminthes

First Edition, 1991
Reprint, 1992, 1999
Second Revised Edition, 2001

ISBN 81-261-0671-9

PRINTED IN INDIA

Published by J.L. Kumar for Anmol Publications Pvt. Ltd., New Delhi - 110 002 and Printed at Mehra Offset Press, Delhi.

Contents

PREFACE

An Introduction to Helminthes is the revised and enlarged second edition of the book. It has been designed to approach the morphology, anatomy, physiology and development of selected type in a simple and lucid style. The approach to the discussion of all the types is very simple so as to impact to the students a clear and vivid understanding. According to the scheme of treatment the important animal types of the phylum have been dealt with first, and efforts have been made to present their elaborate and uptodate account. General characters and classification and brief description of other important types of the phylum have also been dealt with complete, authentic and uptodate account. Further separate chapters on topics of significance and general interest pertaining to the phylum have also been added to make the treatment more elaborate.

It has been the constant endeavour of the authors to furnish maximum substance, keeping in view the limitations of size of the volume. Efforts have been made to condense the matter as far as practicable. The book features both a text and a laboratory guide. It is hoped that this book will not only meet the requirement of Indian students but will also be useful as a guideline to the teachers in their teaching.

There can be no claim to originality except in the manner of treatment and much of the information has been obtained from the books and scientific journals available in different libraries.

The authors express their thanks to their friends and colleagues whose constant inspiration have initiated them to bring out this book.

In this edition all the errors of first edition have been rectified. New information is incorporated in the light of recent developments and latest research. Teachers and scholars can send their criticism and suggestions to further improve the contents.

Authors

1

Helminthes : Characters & Classification

Helminthes constitute a large assemblage of *worms* of comparatively simple and varied organization. Helminthes is derived from the Greek word *helmins* means *worms*. The term worms is not correct because it is applied to elongated invertebrates without appendages and with bilateral symmetary. In 18th Century all the worm-like animals were included in an old phylum-*Vermes* (*Linnaneus*, 1735). But this was a very heterogenous group which included a highly diverse assemblage of forms. In the modern classification this group is divided into several independent phyla such as Platyhelminthes, Aschelminthes, Annelida etc.

The term helminthes is restricted to a few phyla of invertebrate animals, all of which are superficially worm like but they differ markedly in their morphology, life-history and bionomics. The helminthes are restricted to two phyla of Animal Kingdom. These are :

1. Platyhelminthes or flat-worms.
2. Aschelminthes or round-worms.

PHYLUM PLATYHELMINTHES

Term *Platyhelminthes* was proposed by *Gegenbaur* in 1859, means flatworms which refers to their characteristic contour of flattened body. The term derived from two Greek words *-platys* = flat; *helminthes* = worms. The animals show low organization as they are without anus, skeletal, respiratory and circulatory systems. The body is filled with mesenchymal cells which are mesodermal in origin.

General Characters

1. Bilateral symmetrical body, with dorso-ventrally flattened.
2. Body generally shows anterior cephalization that is differentiation of anterior region into head.
3. Body generally worm like either long flat ribbon-shaped or leaf-like. The outline of body bizarre shape in trematodes and segmented tape-like body in cestodes.

4. Most of the flat worms are of small to moderate dimensions to size varies from microscopic to the extreme elongation as much 10 to 15 meters.
5. General colour creamy white but on account of the presence of food the body may acquire colour. Some of the free-living flat worms are brilliantly coloured, often brown, black, gray.
6. Body with various types of attachment in the form of adhesive secretions, suckers, and hooks.
7. Body consists of three embryonic layers viz. ectoderm, mesoderm, endoderm.
8. In parasitic forms, the body is covered with a thick cuticle but free-living turbellarians are clothed with cellular or syncytial epidermis.
9. The exo- and endoskeleton are completely absent therefore the body is generally soft with hooks, spines, spicules and thorns.
10. Body space is filled with a mesenchyma that is parenchymatous tissue, thererfore, flat worms are *acoelomate,* that is without coelom.
11. The digestive system is generally absent in tapeworms and acoela. But in other consists of mouth, pharynx and variously modified intestine.
12. Respiratory and circulatory systems are absent but some of the trematodes have system of tubes the so called *lymphatic system* of uncertain function.
13. The nervous system is of primitive type with missing of nervous tissue as *brain* in anterior-most region with several longitudinal ganglionated cord. Numerous transverse connections occur between the longitudinal cords. So the nervous system is generally ladder like.
14. Parasitic forms are without sensory organs but free-living forms are provided with sensory organs in the form of chemo- and tango-receptors, ciliated pits, statocysts and ocelli or eyes.
15. Excretory system well developed with *proto-nephridial tubules* having flame bulbs or cells. But acoela are without protonephridia.
16. Hermaphrodite, rarely with separate sex. The eggs are generally devoid of yolk but with yolk cells. Varieties of copulatory organs are present. Cross-fertilization and self-fertilization both are common.
17. Development direct i.e. without larval stage or indirect with larval stages. Life cycles complicated with succession of larval stages involving one to three intermediate hosts.
18. Asexual reproduction common among fresh water forms.

19. Parthenogenesis and polyembryony common among trematodes and cestodes.
20. Endogenous and exogenous budding common in tapeworms.

Phylum Platyhelminthes is divided into three classes. The classification followed here is from *Hyman* (1951).

I Class - Turbellaria

II Class - Trematoda

III Class - Cestoda

I. CLASS-TURBELLARIA

Turbellaria is derived from Latin word *'turbella'* means a *stirring*. The animal produce turbulance in water by body cilia. They possess following characters:

1. The members of this class are broad, flat and leaf-like in shape.
2. They lead a free living life in fresh or salt water, in moist soil, some members may be commensal or parasite.
3. Body unsegmented, dorso-ventrally flattened and covered with cilia.
4. The anterior end is differentiated into head.
5. Epidermis cellular or *syncytial* contains rod-like structures called *Rhabidites*.The rhabidites are curved structure capable of discharging a chemical secretion.
6. Epidermis and mesenchyma are provided with unicellular glands of two types (a) *cyanophilous* which secrete adhesive secretion and (b) *eosinophilous*.
7. Turbellarians are abundantly provided with adhesive organs which are of two types (a) *glandulo-epidermal adhesive organs* and (b) *glandulo- muscular adhesive organs*.
8. Alimentary canal incomplete i.e. with single opening, the mouth. Mouth is mid-ventral, sometimes near the centre of body. Intestine is blind or absent.
9. Excretory system with protonephridia.
10. Sense organs consist of tangoreceptors and chaemoreceptors.
11. Mostly bisexual or hermaphrodite or unisexual with very few exceptions.
12. Reproduction by asexual, sexual methods and by regeneration.
13. Cleavage spiral. Development direct without larval stage.

Classification of Turbellaria

The members of this class are found in almost all places. It includes more than 1500 species. Class Turbellalria is divided into 5 orders :

1. Acoela
2. Rhabdocoela
3. Allocoela
4. Tricladida
5. Polycladida

Order 1- Acoela (Gr., *a* = without; *koilos* = cavity)

1. They are minute, delicate turbellarians with a ventral mouth.
2. Mouth and pharynx are present. Pharynx simple when present but usually absent and so the intestine is also absent.
3. Parenchyma syncytial not differentiated into mesoderm and endoderm.
4. Protonephridia absent.
5. Food is taken in solid form and digested by endodermal cells.
6. Gonoducts, gonads and yolk glands are absent.
7. Exclusively marine forms living under stones, algae etc. at the sea bottom. Some live in the intestine of holothurians.

   Examples: *Convoluta, Hoplodisus.*

Order 2- Rhabdocoela (Gr., *rhabdos* = rod; *koilos* = cavity)

1. Small tubellarian usually less than 3mm in length.
2. Mostly free-living, marine or fresh water.

   Some of them inhabit in the respiratory cavities of various fresh water crustaceans, viz. *Temnocephala*.
3. Body is elongated, cylindrical or compressed with simple pharynx and simple unbranched sac-like intestine.
4. Yolk-glands may or may not be present.
5. Male with two to many testes, whereas the female with 1 to 2 ovaries.
6. Nervous system consists of two longitudinal nerve cords. Eyes are present in most of them.

   Order Rhabdocoela is divided into four sub-orders :

Sub-order (i) Notandropora

1. Exclusively fresh water forms.
2. Pharynx simple.
3. Single median protonephridium.
4. Single compact testis mass, penis unarmed.
5. Yolk-gland is absent.
6. Asexual fission with chains of zooids.

   Examples: *Catenula, Rhynchoscolex, Stenostomum.*

Sub-order (ii) Opisthandropora

1. Fresh water or marine forms.
2. Pharynx is simple.
3. Excretory system consists of paired protonephridia.
4. Compact testes, penis armed with a stylet.
5. Yolk glands absent.
6. Asexual reproduction with chains of zooids.

   Examples: *Microstomum, Macrostomum.*

Sub-order (iii) Lecithopora

1. Fresh water, marine or terrestrial, commensal or parasitic.
2. Pharynx bulbose.
3. Protonephridia paired.
4. Separate ovaries and yolk glands.
5. Exclusively sexual reproduction.

   Examples: *Graffilla, Anoplodium, Mesostoma, Gyratrix.*

Sub-order (iv) Temnocephalida

1. Freshwater, ectocommensal forms.
2. Anterior end having 2-12 tentacles.
3. Posterior end with 1 or 2 adhesive discs.
4. Pharynx doliform.
5. Gonopore simple.

   Examples: *Temnocephala, Monodiscus.*

Order 3- Alloecoela (Gr; *alloios* = different; *koilos* = cavity)

1. Mostly marine, few freshwater and brakish water forms. Some are ectoparisite or ectocommensal in habits.
2. The pharynx is bulb-like or plicate.
3. Intestine is straight or branched.
4. Excretory system consists of paired protonephridia with two or three branches and nephridiopores.
5. Nervous system with 3 or 4 pairs of longitudinal nerve cords provided with transverse commissures.
6. Reproductive system consists of numerous testes and a pair of ovaries. Penis papilla mostly present.

   Order Alloecoela is divided into four sub-orders :

Suborder (i) Archiophora

1. Marine animals.

2. Pharynx plicate.
3. Female reproductive system primitive.

   Example: *Proporoplana*

Suborder (ii) Lecithoepitheliata

1. Marine, freshwater or terrestrial forms.
2. Pharynx simple or bulbose.
3. Penis with a cuticular stylet.
4. Female duct simple or none. Typical yolk glands absent, nutritive cells may surround the ova.

   Examples: *Hofstenia, Prorhynchus.*

Suborder (iii) Cumulata

1. Freshwater or marine forms.
2. Pharynx plicate or bulbose.
3. Intestine mostly without diverticulae.
4. Penis unarmed.
5. Separate ovaries and yolk glands or *germovitellaria.*

   Examples: *Hydrolimax, Plagiostomum.*

Suborder (iv) Seriata

1. Mostly marine and a few are freshwater forms.
2. Pharynx plicate.
3. Intestine with lateral diverticula.
4. Statocyst mainly present.
5. Ovaries and yolk glands are separate.

   Examples: *Monocelis, Otoplana.*

Order 4 Tricladida (Gr; *treis* = three; *klados* = branch)

1. Marine, freshwater or terrestrial forms.
2. Large turbellarians, measuring 2-60 cm in length.
3. Pharynx plicate usually backwardly directed.
4. Intestine three-branched sacs.
5. Eyespots usually present.
6. Two ovaries, 2 or numerous testes.
7. Female reproductive system provided with one or two copulatory bursa.
8. Penis with penis papilla, common genital aperture.

   Order Tricladida is divided into three sub-orders :

Sub-order (i) Maricola

1. Exclusively marine forms.
2. A pair of eyes and auricular grooves.
3. Typical penis papilla sometimes armed with stylet.
4. Rounded copulatory bursa is generally present.
5. Only sexual reproduction.

   Examples: *Bdelluora, Ectoplana, Gunda.*

Sub-order (ii) Paludicola

1. Mostly freshwater rarely brackish water forms.
2. Two to many eyes or eyeless.
3. Bursa with bursala canal, anterior to penis.
4. Reproduction mostly asexual.

   Examples: *Dugesia, Kenkia, Dendrocoelum.*

Sub-order (iii) Terricola

1. Terrestrial tropical and subtropical forms.
2. Two or many eyes.
3. Bursa, if present, lies behind the penis.
4. Reproduction may be asexual.

   Examples: *Geoplana, Dolichoplana, Bipalium.*

Order 5 - Polycladida (Gr; *poly* = many; *klados* = branch)

1. Size moderate, from 2-20 mm. Dorsoventrally flat leaf-like body; they are exclusively marine.
2. Intestine with many branches.
3. Eyes and nerve cords are present.
4. There are numerous ovaries and testes.
5. Genital apertures separate.

   It includes two suborders:

Sub-order (i) Acotylea

1. Pharynx usually vertical curtain like.
2. No suckers are present behind the female pore.
3. Tentacles nuchal type.
4. Eyes concentrated at anterior margin, not in pairs.

   Examples: *Stylochus, Notoplana.*

Sub-order (ii) Cotylea

1. Sucker present behind the female pore.

2. Pharynx tubular.
3. A pair of marginal tentacles bearing eyes or with a cluster of eyes in place of tentacles.

   Examples: *Thysasnozoon, Yungia*

CLASS-II. TREMATODA

1. They are ecto- or endoparasites and commonly called *'flukes'*.
2. Body is usually unsegmented, dorso-ventrally flattened and leaf-like.
3. Size varies from 0.5 mm to 1 or several cm. long.
4. Body is covered by thick cuticle and no cilia.
5. Suckers and sometimes hooks area present. Oral sucker present at anterior end for feeding, ventral sucker for attachment.
6. Epidermis absent. *Rhabdites* are absent.
7. Digestive system consists of mouth, muscular pharynx and intestine usually with two main branches. Anus is absent.
8. In some cases there is a system of parenchymal vessels showing the presence of primitive circulatory system.
9. Excretory organs consist of protonephridia.
10. There are three pairs of longitudinal nerve cords. Sense organs are poorly developed.
11. Bisexual or hermaphrodite except blood fluke (*Schistosoma*).
12. Single ovary and many testes. Cross fertilization takes place. Development direct or indirect.
13. Life-cycle simple or complicated with more than one hosts.

    There are about 6000 species in class Trematoda. This class is divided into three orders:

    Order 1 - Monogenea

    Order 2 - Aspidobothria

    Order 3 - Digenia

Order 1 - Monogenea

1. They are ectoplarasite mostly in cold-blooded animals (fishes).
2. The life-cycle is simple and passed in one host only, *monogenetic*.
3. Oral sucker absent, when present is poorly developed.
4. A large disc-shaped structure, the *opisthaptor*, is found on the posterior end. It contains hooks and is lined with chitin.
5. Paired excretory pores are present in the anterior part of the body.
6. Male and female genital pores are separate.

7. One or two vaginae are present. The uterus is small and contains few eggs.
8. Development is indirect through a larval stage called onchomiracidium.

Order Monogenea is divided in two suborders:

Sub-order (i) Monopisthocotylea

1. Opisthaptors are present without suckers but with hooks.
2. Genito-intestinal tract is absent.
3. Posterior adhesive organ is single or divided by septa.
4. Eye-spots usually present.

Examples: *Gyrodactylus, Monocotyle, Heterocotyle, Amphibdella*

Sub-order (ii) Polyopisthocotyle

1. Opisthaptor contains many suckers and also hooks.
2. Oral suckers two, opening into oral cavity.
3. Genito-intestinal tract present.
4. Eye-spots are absent.

Examples: *Polystoma, Diplozoon, Microcotyle.*

Order 2 - Aspidobothrea

1. Endoparasite in the digestive tract of fishes and reptiles.
2. Oral sucker is absent.
3. The ventral sucker is divided by septa into many small parts. The hooks are absent in it.
4. Single excretory pore.
5. Only one testis is present.
6. Development direct, without alternation of hosts.

Order Aspidobothrea is divided into two families:

Family (a) Aspidogastridae

1. Sucker enormous circular, oval or elongated, divided by septa into one, three or four longitudinal rows of depressions termed alveloli.
2. Protonepliridia have separate bladders and separate or a common nephridiopore situated near the posterior end.
3. Testis single.
4. Laurer' canal may be present.
5. Endoparasite in the mantle, pericardial and renal cavities of clams and snail and in digestive system of fish and turtle.

Examples: *Aspidogaster, Cotylaspis, Cotylogaster*

Family (b) Stichocotylidae

1. It consists of only one genus which lives in bile passage or spiral valve of fishes.
2. The life cycle is complex one, involving at least two hosts.

Example: *Stichocotyle.*

Order 3 - Digenea

1. They are endoparasite in invertebrate and vertebrates.
2. Two to four hosts in the life-cycle.
3. Two suckers, without hooks, mouth within anterior sucker.
4. There is only one excretory pore present in the posterior region of the body.
5. There is a common gonopore for male and female reproductive systems. Vagina is absent.
6. Uterus is long branched with many eggs.
7. Their larval forms reproduce asexually before metamorphosis.

Order Digenea is divided into many families; some important ones are as :

Family (a) Bucephalidae

1. They are commonly called gastrostomes. The mouth is located in the middle of the body.
2. Anterior end bears a simple or lobate oral sucker or has a generally lobed hood-like extension.
3. Intestine sacciform.
4. Acetabulum absent.
5. Genital pore posterior.
6. The cercaria larvae known as oxhead, have a pair of long tails.
7. The primary host is fish and secondary host is clams.

Examples: *Bucephalus, Bucephalopsis.*

Family (b) Paramphistomidae

1. Body thick and fleshy with a powerful posterior sucker.
2. Cirrus pouch is present.
3. Oral sucker is absent.
4. Gonopore, situated in the anterior body half and provided within some genera with a genital sucker.
5. Nephridiopore dorsal in front of acetabulum.
6. Uterus long having ascending limb. Vitelline glands follicular.

7. These are the endoparasites of gut and bile ducts.
   Examples: *Paramphistomum, Gastrothylax.*

Family (c) Opisthorchidae

1. These are distomes of small to large size.
2. They live in the bile passage of fish-eating reptiles, birds and mammals.
3. Body flat and semitransparent.
4. Excretory bladder 'V' shaped.
5. Testes posterior. Genital pore in front of acetabulum.
   Examples: *Opisthorchis, Amphimerus, Metorchis.*

Family (d) Fasciolidae

1. Commonly known as *liver flukes.*
2. Body leaf-like and acetabulum near the oral sucker.
3. Gonopore just in front of the acetabulum.
4. Testes, ovary, vitelline glands all are branched.
   Examples: *Fasciola hepatica, Fasciplopsis buski.*

Family (e) Echinostomidae

1. Body slender, elongated with spiny or scaly cuticle.
2. Acetabulum large behind the oral sucker.
3. Collar of spines is present in cercarial and metacercarial stages.
4. It is commonly found in the intestine of birds.
   Examples: *Echinostoma, Echinoparyphium, Parorchis, Himasthla.*

Family (f) Schistosomatidae

1. These are *blood flukes* that inhabit the hepatic portal system and pelvic veins of birds and mammals.
2. Sexes are separate.
3. Female lodges in a *gynaecophoric canal* of male.
4. Testes numerous and ovary tubular.
   Example : *Schistosoma.*

Besides above families the others are:

Family - Zoognidae.	Examples : *Zoogonus, Nassa*
Family - Allocregdidae.	Examples : *Bunodera, Plagioporus.*
Family - Azygiidae.	Examples : *Azygia, Otodistomum.*
Family - Hemiuridae.	Examples : *Halipegus, Derogenes.*
Family - Didymozoonidae.	Example : *Didymocystis.*
Family - Gorgoderidae.	Examples : *Gorgoderina, Gorgodera.*

Family - Plagiorchidae. Examples : *Plagiorchis*, *Styphlodora*.
Family - Dicrocoelidae. Examples : *Dicrocoelium*, *Eurytrema*.
Family - Heterophyidae. Examples : *Stictodera*, *Heterophyes*.
Family - Microphallidae. Examples : *Spelotrema*, *Maritrema*.
Family - Troglotrematidae. Examples : *Paragonimus*, *Troglotrema*.
Family - Cyclocoelidae. Examples : *Cyclocoelus*, *Typhlocoelum*.
Family - Clinostomatidae. Example : *Clinostomum*.
Family - Notocotylidae. Examples : *Notocotylus*, *Catatropis*.
Family - Strigeidae. Examples : *Ophiososma*, *Cyanthocotyle*.
Family - Brachylaimidae. Example : *Brachylaima*.
Family - Sanguinicolidae. Example : *Sanguinicola*.
Family - Spirorchidae. Examples : *Spirorchis*, *Hapalorhynchus*.
Family - Heronimidae. Example : *Heronimus*.

CLASS III CESTODA

1. They are found as endoparasites in the intestine of vertebrates and are called *tapeworms*.
2. Body is flat, elongate and ribbon-like, measures from 1 mm to 10 metres in length.
3. Body is without epidermis, rhabdites, cilia but a cuticle is present.
4. Body is usually divided into a scolex or head, neck and few to many proglottids.
5. Scolex is provided with hooks or suckers or both.
6. Scolex is followed by neck which forms new '*proglottids*' by transverse budding called *strobilization*.
7. Mouth and digestive tract are absent.
8. Excretory system consists of protonephridia with flame-cells.
9. Nervous system consists of ganglionated nerve ring in scolex and two pairs of lateral nerve cords in proglottids.
10. Each proglottid contains one or two sets of male and female reproductive organs. Thus they are hermaphrodite.
11. Several testes in the form of follicles and single-lobed ovary.
12. Uterus presents a great variety of structures.
13. Fertilization internal and self.
14. Life-cycle is complicated with hooked embryo, and is passed in more than one host.

Class Cestoda has about 3400 species. The class is divided into two sub-classes :

Sub-class (A) Cestodaria or Monozoa

Sub-class (B) Eucestoda or Merozoa

Sub-Class (A) Cestodaria

1. Endoparasites in the intestine and coelomic cavities of fishes and reptiles.
2. Body is formed by one segment and without a scolex.
3. No alimentary canal.
4. Single set of reproductive organs.
5. Larva with ten hooks, known as *pycophore.*

Sub-class Cestodaria is divided into three orders :

Order 1 - Amphilinidea

1. They are endoparasites in the body cavity of ganoid fishes.
2. Frontal gland and protrusible pharynx are present in the anterior part of the body.
3. Suckers are absent.
4. A proboscis is present.
5. Male and female pores present in the posterior part of body.
6. Uterus is coiled and opens near the anterior end.

Example: *Amphilina*

Order 2 - Gyrocotylidea

1. They are endoparasites of chimaeroid fishes.
2. Anterior end has cup-like sucker.
3. Eversible proboscis at anterior end.
4. Male and female pores are situated at the anterior half of the body.
5. Uterus not coiled.

Example: *Gyrocotyle.*

Order 3 - Biporophyllidea

1. Endoparasites in the intestine of sharks.
2. Proboscis on anterior end.
3. Male and female gonopores open in a common atrium.

Example : Biporophyllaeus.

Sub-Class (B) Eucestoda

1. Elongated, flat body is divided into many proglottids.
2. Body is divisible into scolex, neck and strobila.
3. Scolex with suckers, spines and hooks for attachment to host.

4. Each proglottid with male and female reproductive organs.
5. Larva with six hooks.

Sub-class Eucestoda is divided into following orders:

Order 1 - Tetraphyllidea

1. Endoparasite, exclusively in the intestine of elasmobranch fishes.
2. Scolex provided with four bothria (sessile suckers) often with hooks.
3. Testes lie infront of ovaries.
4. Cirrus armed with hairs, spines or hooks.
5. Common genital atrium marginal.

Examples: *Phyllobothrium, Acanthobothrium.*

Order 2 - Lecanicephaloidea

1. Intestinal parasites of elasmobranch fishes.
2. Scolex divided by a transverse groove in two parts upper with disc and lower with four suckers.
3. Vitellaria in two lateral bands.

Examples: *Lecanicephalum, Polypocephalus.*

Order 3 - Proteocephalidea

1. These are endoparasites of freshwater fishes, amphibians and reptiles.
2. Scolex mobile with four lateral suckers, an apical sucker in some.
3. Ovary bilobed. Uterus with many branches and pores.
4. Vitellaria scattered.
5. Common genital atrium marginal.

Examples: *Proteocephalus, Ophiotaenia.*

Order 4 - Diphyllidea

1. Parasitic in the intestine of elasmobranch fishes.
2. Proglottids only 20 or less.
3. Scolex with two bothria and a spiny head stalk.
4. Vitellaria scattered.

Example: *Echinobothrium.*

Order 5 - Trypanorhyncha

1. Parasitic in the spiral valve of digestive tract of elasmobanch fishes.
2. Strobila is segmented and is of moderate size.
3. Scolex with two or four sessile bothria and four protrusible spiny proboscis.
4. Testes extend beyond ovary posteriorly.

Examples: *Haplobothrium, Tetrarhynchus.*

Order 6 - Pseudophyllidea

1. Endoparasites in the intestine of fishes, birds and mammals.
2. The larva lives in crustaceans.
3. Generally six shallow groove-like suckers and hookless bothria are found on the scolex.
4. Proglottids may be absent or present.
5. Neck is not developed.
6. Many yolk glands are present.

   Examples: *Amphicotyle, Dibothriocephalus.*

Order 7 - Nippotaeniidea

1. Endoparasite in the intestine of Japanese freshwater fishes.
2. Scolex with single terminal sucker.
3. Proglottids few, each with one set of genitalia.
4. Vitellaria compact.

   Examples: *Nippotaenia, Amurotaenia.*

Order 8 - Taenioidea

1. Endoparasites in the intestine of birds and mammals.
2. Elongated body, with proglottids.
3. Scolex bears four suckers often with an apical rostellum armed with hooks.
4. Excretory system with four longitudinal vessels and flame cells.
5. Complete sets of reproductive organs in each proglottids. Single compact yolk gland.

   Examples: *Taenia, Echinococcus.*

Order 9 - Aporidea

1. It is endoparasite in swan (birds).
2. Scolex with four suckers, rostellum armed.
3. No segmentation, ootype, yolk glands and sex-ducts or pores.

   Examples: *Nematoparataenia, Gastrotaenia.*

POLYCHOERUS

Phylum	-	Platyhelminthes
Class	-	Turbellaria
Order	-	Acoela
Genus	-	*Polychoerus*

1. It is found in littoral zone of ocean shores in colder winter.

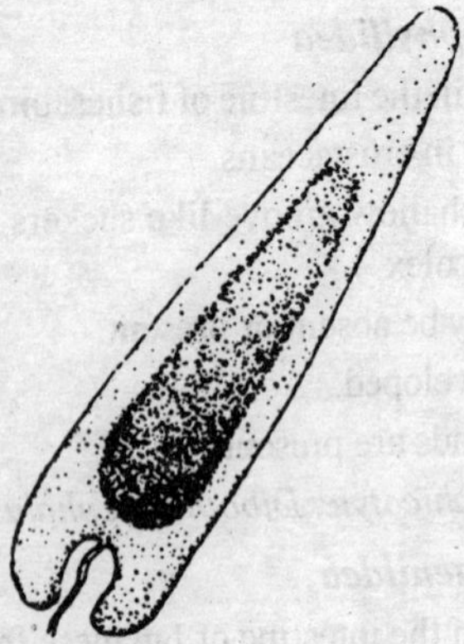

Fig.1.1. *Polychoerus.*

2. It is bright orange or orange red in colour.
3. Body flat, oval, changeable shape. The anterior end is round while posterior end is not ched bearing tail filament.
4. Entire body is covered with cilia.
5. Statocyst is present in median line.
6. Digestive tract is absent.
7. Definite gonads are wanting. Yolk glands are absent. Definite ovduct absent.
8. The seminal bursa opens ventrally by female gonopore.
9. Penis is present.

CONVOLUTA

Phylum	-	Platyhelminthes
Class	-	Turbellaria
Order	-	Acoela
Genus	-	*Convoluta*

1. *Convoluta* is found in sea water.
2. Living under stones, among algae and most commonly on the bottom mud.
3. Small worm with sides of the body are held curved ventrally without any projections.
4. The body has pigment cells.
5. The anterior or cephalic end has two eyes. a statocyst and cluster of frontal glands.

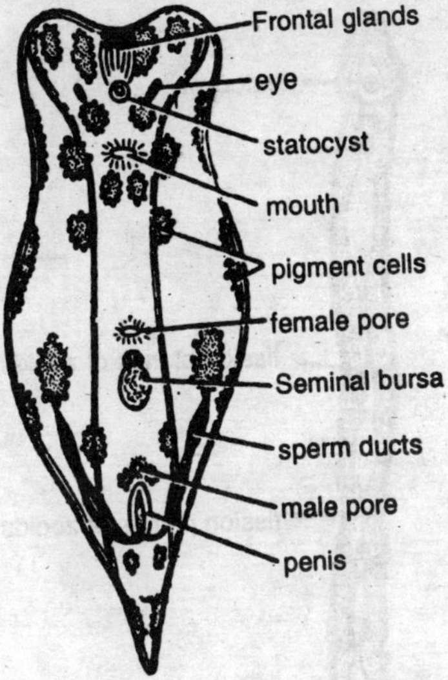

Fig. 1.2. *Convoluta.*

6. The mouth is located near the anterior half of the ventral surface.
7. A digestive cavity is lacking.
8. The excretory system is completly absent.
9. Hermaphrodite but protandry is common.
10. The eggs are enclosed in a thin gelatinous capsule and are struck to some objects.
11. The eggs are developed directly into young worms that escape from the jelly.

CATENULA

Phylum	-	Platyhelminthes
Class	-	Turbellaria
Order	-	Rhabdocoela
Suborder	-	Notandropora
Genus	-	*Catenula*

1. It is a freshewater Turbellaria.
2. The pharynx is simple.
3. Single median protonephridium is present.
4. Nervous system consists of four pairs of longitudinal nerve cords and a statocyst.

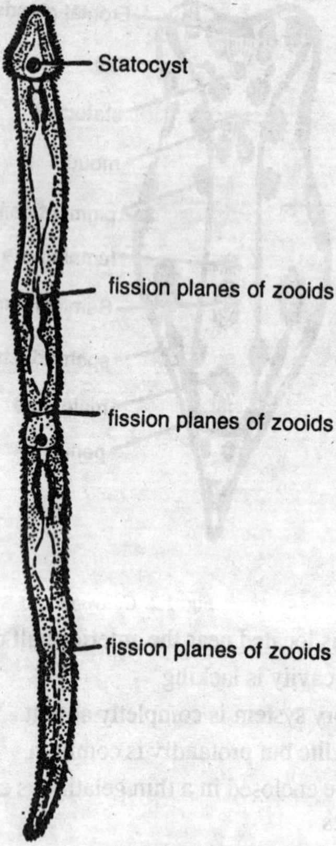

Fig. 1.3. *Catenula.*

5. Asexual reproduction by fission with chains of zooids.
6. No sexual reproduction.

MICROSTOMUM

Phylum	-	Platyhelminthes
Class	-	Turbellaria
Order	-	Rhabdocoela
Suborder	-	Opisthandropora
Genus	-	*Microstomum*

1. It is a microscopic, freshwater form.
2. It swims by means of cilia present on the body surface.

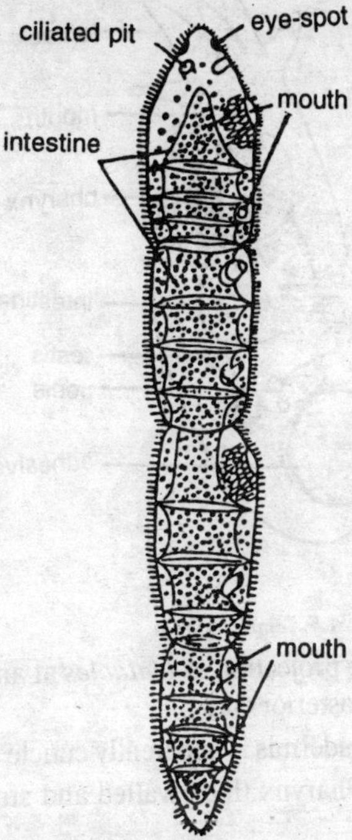

Fig. 1.4. *Microstomum.*

3. Its epidermis contains nematocysts derived from hydroid food.
4. It is found that *Microstomum* attacks *Hydra* only when it needs of nematocysts.
5. The digestive cavity is simple, unbranched and straight.
6. It reproduces asexually by repeated transverse fission forming a linear chain of 4, 8 or 16 zooids. Each zooid separates to become an individual.

TEMNOCEPHALA

Phylum	-	Platyhelminthes
Class	-	Turbellaria
Order	-	Rhabdocoela
Suborder	-	Temnocephalida
Genus	-	*Temnocephala*

1. It is found as an ectocommensal on cray-fishes. prawns and other crustaceans.

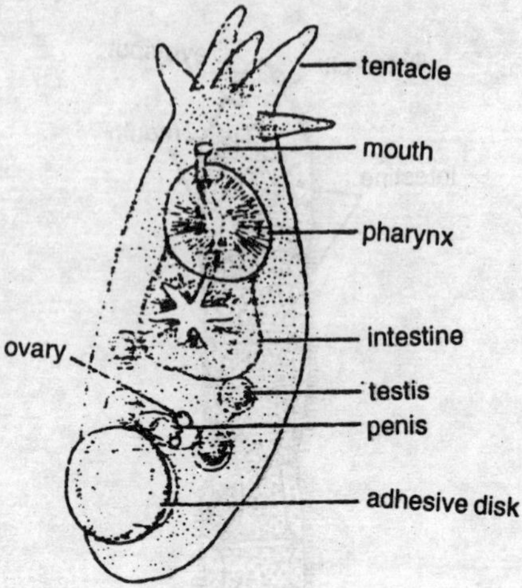

Fig. 1.5 *Temnocephala.*

2. It contains 12 finger-like projection or *tentacles* at anterior end and an *adhesive disc* at the posterior end.
3. The border of syncytial epidermis is apparently cuticle devoid of cilia.
4. Mouth anterio-ventral. Pharynx thick walled and simple lobulated intestine present.
5. It feeds upon diatoms, rotifers and other small animals.
6. A pair of nephridiopores is present on the dorsal surface of the anterior part of the body.
7. Cerebral ganglion bears two dorsal eyes.
8. There are four testes and single ovary. Their gonoducts leading into a common genital atrium.

HOFSTENIA

Phylum	-	Platyhelminthes
Class	-	Turbellaria
Order	-	Alloecoela
Suborder	-	Lecithoepitheliata
Genus	-	*Hofstenia*
Species	-	*atroviridis*

1. It is a marine animal found in Japan.
2. The pharynx is very long, muscular and simple.
3. The muscle fibres run through epidermis, parallel to the surface and appear to be part of epidermis.

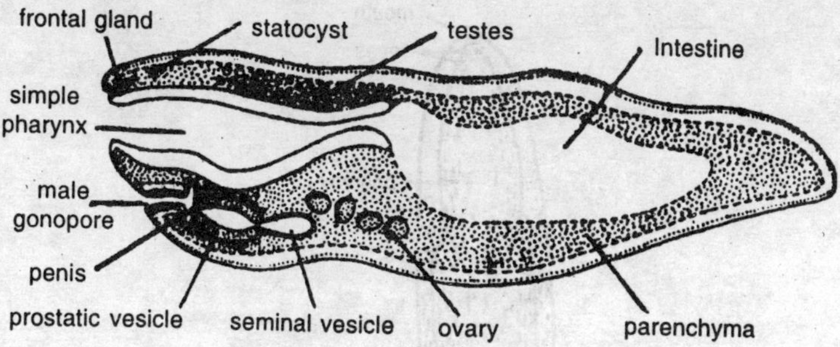

Fig. 1.6. *Hofstenia atroviridis*

4. At the anterior end besides the main nervous system, there is a delicate plexus.
5. Frontal gland and a statocyst are present.
6. Germovitellaria consist of loose clusters. Each cluster having one oocyte surrounded by nutritive cells. The yolk glands are absent.
7. Bisexual. There is a simple female duct and typical male apparatus, armed with a penis-stylet is present.

PLAGIOSTOMUM

Phylum	-	Platyhelminthes
Class	-	Turbellaria
Order	-	Allocoela
Suborder	-	Cumulata
Genus	-	*Plasgiostomum*

1. It is present in both freshwater and marine.
2. Mouth is anterior.
3. Pharynx is bulbose. Intestine with short diverticula.
4. Gonopore posterior.
5. Sperm is characteristic in having wing-like lateral expansions.

BIPALIUM

Phylum	-	Platyhelminthes
Class	-	Turbellaria
Order	-	Tricladida
Suborder	-	Terricola
Genus	-	*Bipalium*

1. *Bipalium* is a long terrestrial turbellarian living the humid soil on the floor of tropical forests.

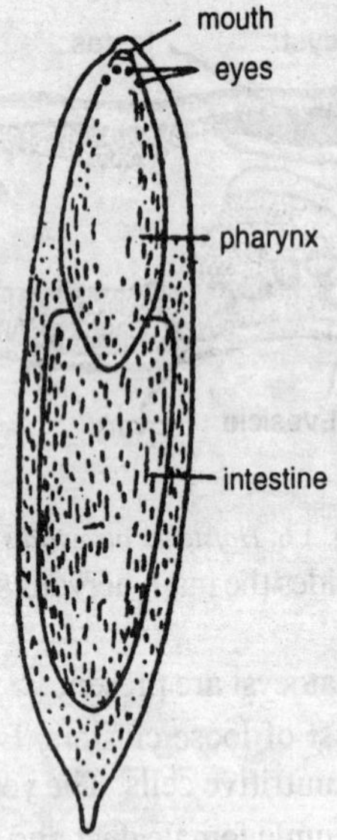

Fig. 1.7. *Plagiostomum*

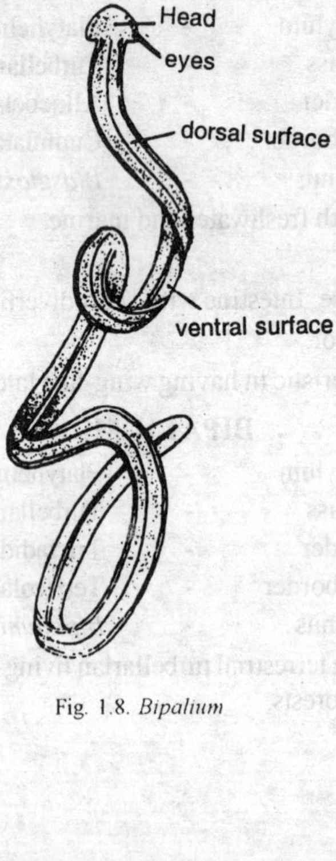

Fig. 1.8. *Bipalium*

2. It is cosmopolitan in distribution.
3. The animal measures 20-50 cm in length.
4. It consists of an expanded lunate head and cylindrical long body.
5. Numerous eyes are present on the margin of the head and sides of body.
6. There are stripes on the dorsal surface and a creeping sole on the ventral surface.
7. The stripes are purple, black, yellow, olive and grey in colour.
8. Reproduction generally asexual. It never becomes sexual in temperate climate and it propogates by fragmentation.
9. *Bipalium adventium* breeds sexually.

GUNDA

Phylum	-	Platyhelminthes
Class	-	Turbellaria
Order	-	Tricladida
Suborder	-	Maricola
Genus	-	*Gunda* (*Procerodes*)

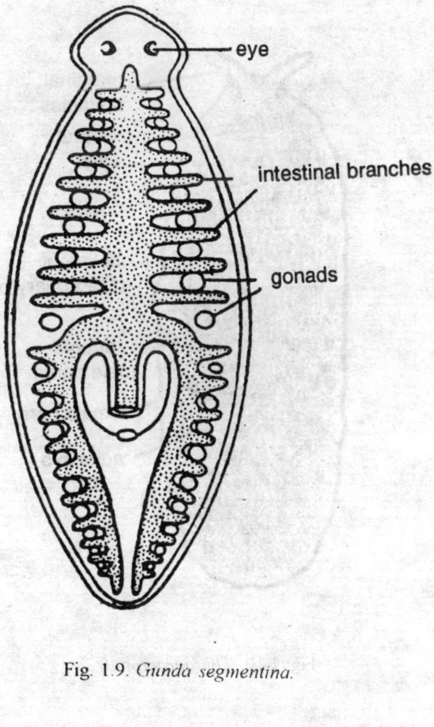

Fig. 1.9. *Gunda segmentina.*

1. It is a marine animal.
2. The auricle and a pair of eye or ocelli are present in the anterior end.
3. Intestine has regularly intestinal diverticulae. The gonads alternate with the diverticulae giving the appearance of segmentation.
4. Incapable of sexual reproduction.
5. Long used the feature of intestine as a classical example in support the gonocoel formation by *Haeckel*.

THYSANOZOON

Phylum	-	Platyhelminthes
Class	-	Turbellaria
Order	-	Polycladida
Suborder	-	Cotylea
Genus	-	*Thysanozoon*

1. It is found in a less cold water of sea.
2. It has an oval body with dorsal surface beset with papillae, each receiving a branch from intestine.
3. Anterior end bears a pair of marginal tentacles and numerous cerebral eyes.

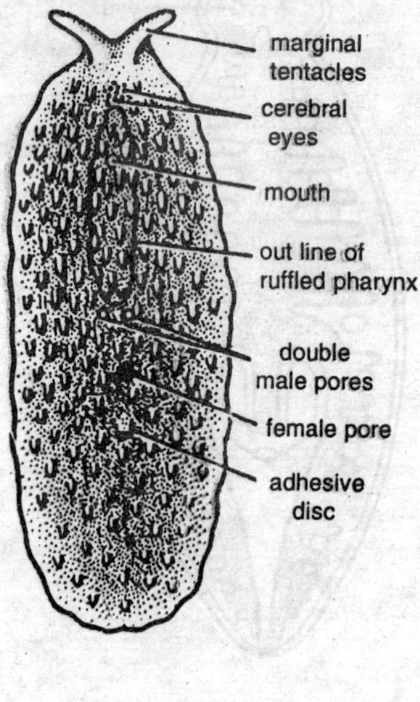

Fig. 1.10. *Thysanozoon.*

4. Pharynx is tubular.
5. Glandulo-muscular adhesive organ is present on the dorsal surface behind the female gonopores.
6. A single female genital pore lies behind the male genital pores.
7. Seminal bursa is absent.

ASPIDOGASTER

Phylum	-	Platyhelminthes
Class	-	Trematoda
Order	-	Aspidobothria
Genus	-	*Aspidogaster*

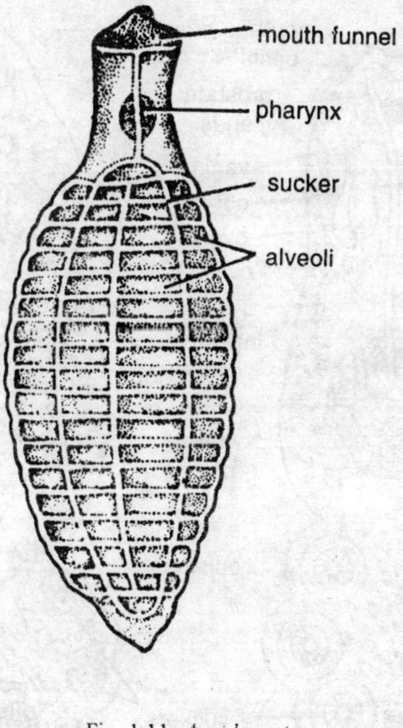

Fig. 1.11. *Aspidogaster.*

1. It is an endoparasite in the pericardial and renal cavities of fresh water mussel and in the gut of fishes and turtles.
2. Body is elongated and dorsoventrally flattened with anterior narrow end.
3. The narrow anterior end has a subterminal mouth which is devoid of oral sucker.
4. A large sucker appears which occupies great space in the posterior

region of body. The sucker is sub-divided into four longitudinal rows of sucking cups or alveoli.

5. Gut is simple and straight.
6. Excretory system consists of protonephridia with excretory bladders.
7. Bisexual life-cycle without change of host.

POLYSTOMUM

Phylum	-	Platyhelminthes
Class	-	Trematoda
Order	-	Monogenea
Suborder	-	Polyopisthocotylea
Genus	-	*Polystomum*

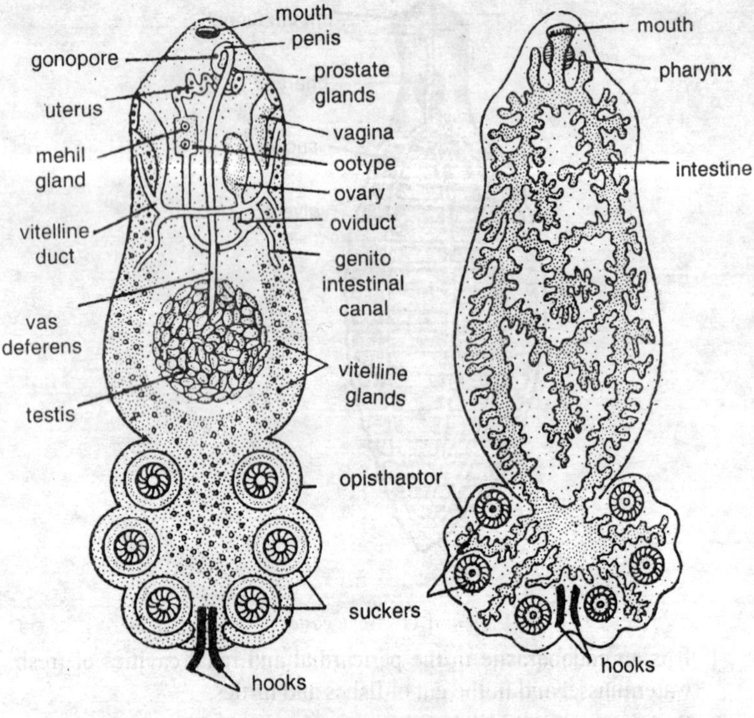

Fig. 1.12. *Polystomum*

1. It is an endoparasite in the urinary bladdar of frogs. and turtles.
2. Body is leaf-like and dorso-ventally flattened.
3. The posterior region has a disc-shaped opisthaptor with six cup-like suckers and 2 to 4 large chitinous hooks.

4. Mouth is surrounded by an oral sucker.
5. Intestine is bifurcated.
6. Hermaphrodite with single ovary and numerous testes.
7. Breeding season starts in spring. Eggs are produced in the urinary bladder of frog and discharged into the water alongwith the urine of the host.
8. Hatching takes place after 4 to 5 weeks and larva attaches to the gills of tadpole and later develop into adult in the alimentary canal of metamorphosed tadpole.

DIPLOZOON

Phylum	-	Platyhelminthes
Class	-	Trematoda
Order	-	Monogenea
Suborder	-	Polyopisthocotylea
Genus	-	*Diplozoon*

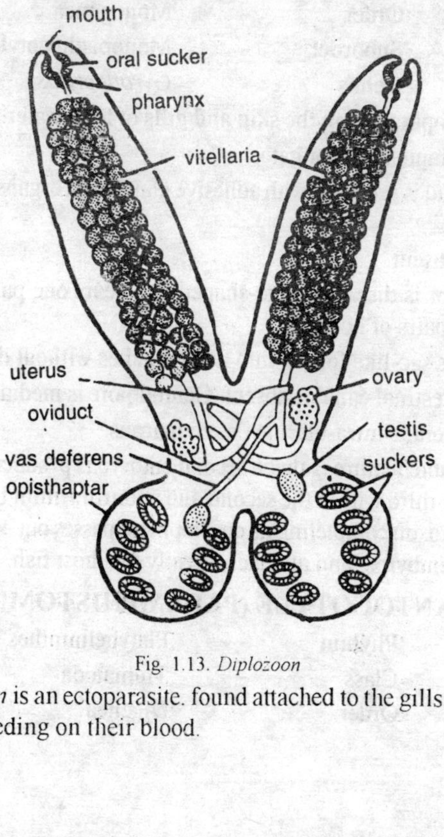

Fig. 1.13. *Diplozoon*

1. *Diplozoon* is an ectoparasite. found attached to the gills of freshwater fishes. feeding on their blood.

2. Two individuals are always found attached permanently together in the form of 'X'.
3. Opisthaptor bears eight suckers arranged in two rows of four each.
4. Digestive system consists of mouth surrounded by oral sucker, pharynx and intestine. Intestine is not bifurcated.
5. Reproductive system consists of a testis and ovary along with their usual ducts.
6. The sperm duct, uterus and vagina of one individual cross with those of other and thus ensures cross fertilization. Uterus surrounded by *Mehlis' glands* and contains one egg.
7. Egg differenciates into diporpa larva, two of which unite and metamorphosed into adult individual. Larva has two eyes, digestive system and a pair of suckers.

GYRODACTYLUS

Phylum	-	Platyhelminthes
Class	-	Trematoda
Order	-	Monogenea
Suborder	-	Monopisthocotylea
Genus	-	*Gyrodactylus*

1. It is an ectoparasite on the skin and gills of freshwater fishes.
2. Body is minute and elongate.
3. Anterior end is provided with adhesive glands and organs. Oral sucker is absent.
4. Eyes are absent.
5. Opisthaptor is disc or wedge-shaped and bears one pair of anchors and eight pairs of hooklets.
6. Intestine is sac-like forked into two branches without diverticula.
7. Genito-intestinal canal is absent. Genital pore is median.
8. Development is intra-uterine i.e. viviparous.
9. A remarkable feature is that a second embryo is produced within the first one, a third within the second and a fourth within the third. The first embryo. on completing its development passes out, still enclosing the other embyros, and attaches directly to a host fish.

GIGANTOCOTYLE (PARAMPHISTOMUM)

Phylum	-	Platyhelminthes
Class	-	Trematoda
Order	-	Digenea

Family	-	Paramphistomidae
Genus	-	*Gigantocotyle*

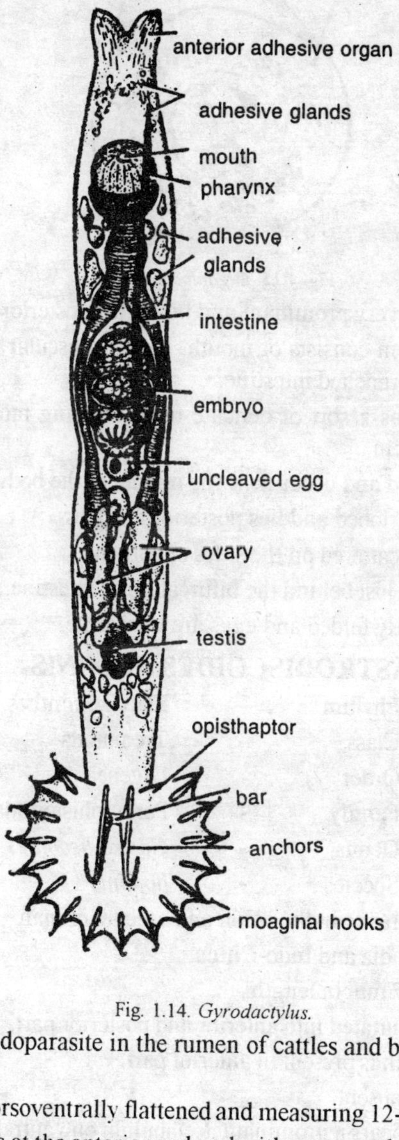

Fig. 1.14. *Gyrodactylus.*

1. It is an endoparasite in the rumen of cattles and bile-duct of sheep, goat etc.
2. Body is dorsoventrally flattened and measuring 12-14 mm in length.
3. Mouth lies at the anterior end and without an oral sucker.

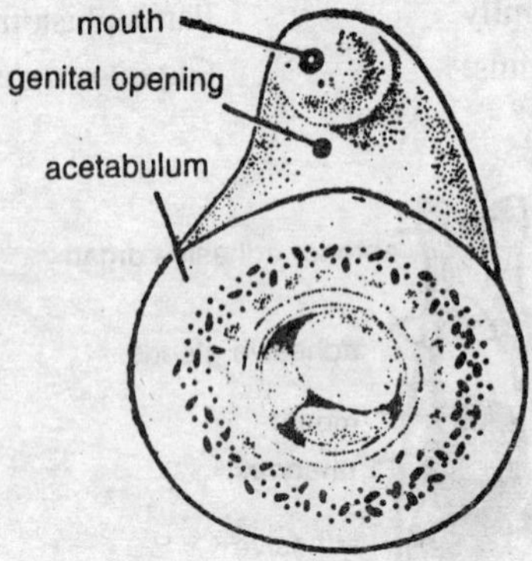

Fig. 1.15. *Gigantocotyle.*

4. Acetabulum is very prominant and lies at the posterior end.
5. Digestive system consists of mouth, highly muscular pharynx and bifurcated, unbranched intestine.
6. Parenchyma has a sort of delicate tubes forming much branched lymphatic system.
7. Testes are lobed and occur near the middle of the body.
8. Ovary is small, lobed and lies posterior to testes.
9. Vitellaria are scattered on the sides of animal.
10. Genital pore is just behind the bifurcation of intestine.
11. Uterus is slightly folded and eggs are very large.

GASTRODISCOIDES HOMINIS

Phylum	-	Platyhelminthes
Class	-	Trematoda
Order	-	Digenea
Family	-	Paramphistomidae
Genus	-	*Gastrodiscoides*
Species	-	*hominis*

1. It is an endoparasite in the colon and caecum of man.
2. It is found in India and Indo-China.
3. It measures 5-7 mm in length.
4. Body is differentiated into anterior and posterior part.
5. Terminal mouth is present in anterior part.
6. Oral sucker is absent.
7. Posterior part bears a prominant acetabulum on ventral side.

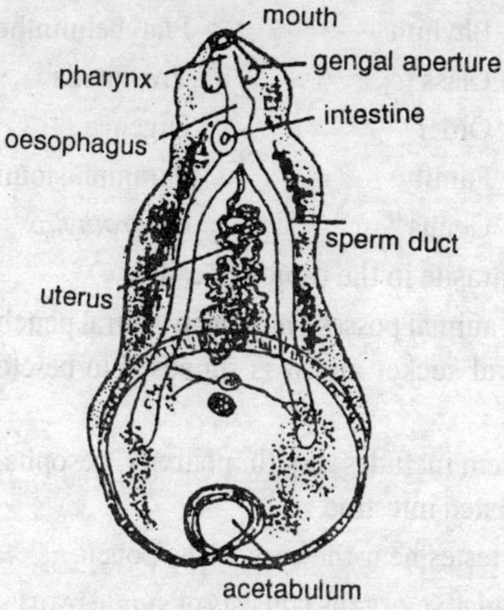

Fig. 1.16. *Gastrodiscoides.*

8. Gonopore is in the anterior half of the body on ventral side.
9. Excretory pore lies on the dorsal side infront of acetabulum.
10. The secondary host is snail.
11. It causes inflammation and diarrhoea.

GASTROTHYLAX

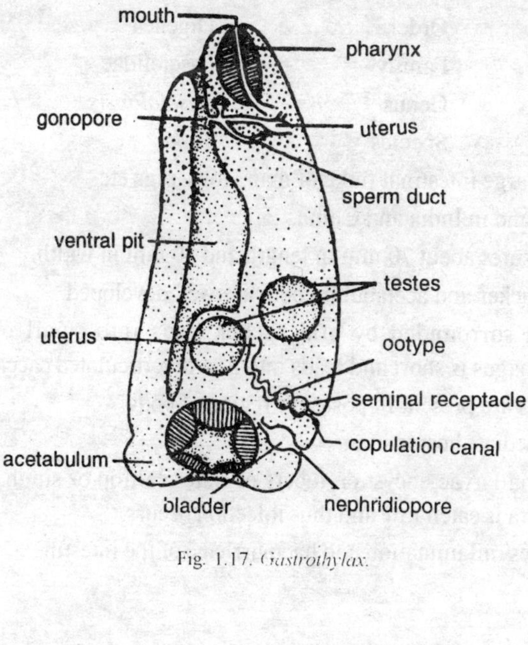

Fig. 1.17. *Gastrothylax.*

Phylum	-	Platyhelminthes
Class	-	Trematoda
Order	-	Digenea
Family	-	Paramphistomidae
Genus	-	*Gastrothylax*

1. It is an endoparasite in the rumen of cattles.
2. It is elongated animal possessing a deep ventral pouch which extends upto the ventral sucker and it is supposed to be elongated genital chamber.
3. Digestive system includes mouth, pharynx, oesophagus and simple non-diverticulated intestine.
4. Two rounded testes near the base of the pouch.
5. Female reproductive organs consists of single ovary, uterus, ootype, vagina and seminal recepticle.
6. Single excretory pore is present.
7. Posterior end is divided with large acetabulum or adhesive sucker.

FASCIOLOPSIS BUSKI

Phylum	-	Platyhelminthes
Class	-	Trematoda
Order	-	Digenea
Family	-	Fasciolidae
Genus	-	*Fasciolopsis*
Species	-	*buski*

1. It is a large intestinal fluke of man, dogs, pigs etc.
2. It is found in India and China.
3. It measures about 70 mm in length and 20 mm in width.
4. Oral sucker and acetabulum both are well developed.
5. Mouth surrounded by oral sucker leads into small pharynx. Oesophagus is short and leads into non-diverticulated caeca.
6. Gonads are present in posterior region of body.
7. Intermediate host is snail.
8. Cercaria larvae encyst on tubers of water caltrop or singhara. The singhara is eaten raw and thus infection occurs.
9. It causes inflammation and haemorrhage of the intestine.

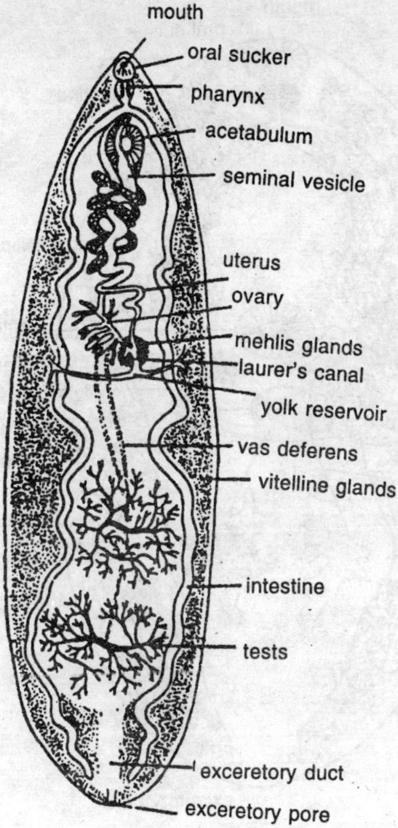

Fig. 1.18. *Fasciolopsis buski.*

PARAGONIMUS WESTERMANNI

Phylum	-	Platyhelminthes
Class	-	Trematoda
Order	-	Digenea
Family	-	Trglotrematidae
Genus	-	*Paragonimus*
Species	-	*westermanni*

1. *Paragonimus westermanni or lung fluke* is parasite in the lungs of mammal including man.
2. The body of the fluke is oval with well developed oral sucker surrounding the mouth and a ventral sucker in the centre of the body.
3. The internal organs are very much like that of *Fusciola hepatica.*

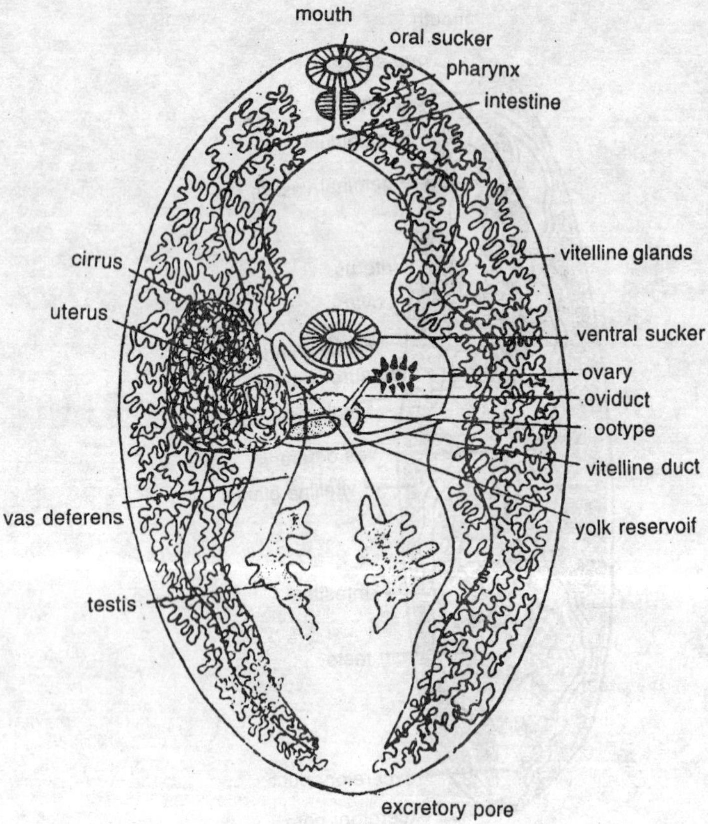

Fig. 1.19. *Paragonimus.*

4. The cyst like pockets are formed around the parasite which rupture and liberate the eggs into bronchial tubes to be executed with sputum.
5. The miracidia live only few hours after hatching.
6. The cercariae are long, have small knob-like tail and spiny cuticle.
7. They do not swim but creep in leech-like manner.
8. They pierce the cuticle of cray-fish where they become encysted inside the tissues and gradually develop into mature infective metacercariae.
9. The infection of the new host usually takes place by eating infected crab or cray-fish.
10. They cause or help in tuberculosis, cough and blood stained sputum.

AMPHILINA

Phylum	-	Platyhelminthes
Class	-	Cestoda
Subclass	-	Cestodaria
Order	-	Amphilinidea
Genus	-	*Amphilina*

1. It is an endoparasite found in the coelom of fish, *Acipenser*.
2. Body flat and leaf-like. Scolex is absent.
3. Anterior end is provided with weakly develop protrusible proboscis.
4. Testes are scattered and cirrus is armed.
5. Vagina lies behind the ovary and opens on the side. Uterus is coiled.
6. The intermediate host is a crustacean, *Gammarus*, which is eaten by *Acipenser*.

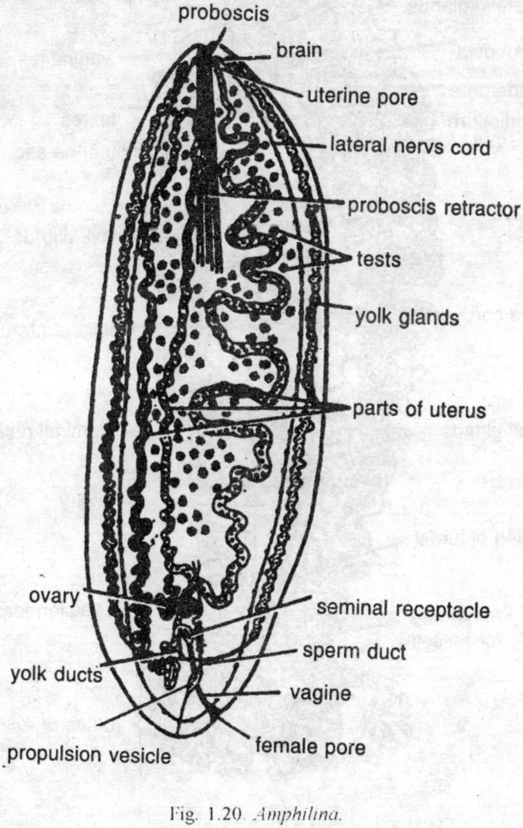

Fig. 1.20. *Amphilina.*

GYROCOTYLE

Phylum	-	Platyhelminthes
Class	-	Cestoda
Subclass	-	Cestodaria
Order	-	Gyrocotylidea
Genus	-	*Gyrocotyle*

1. These are endoparasite in the intestine of chimaeroid fishes.
2. Body is elongated and dorsoventrally flattened.

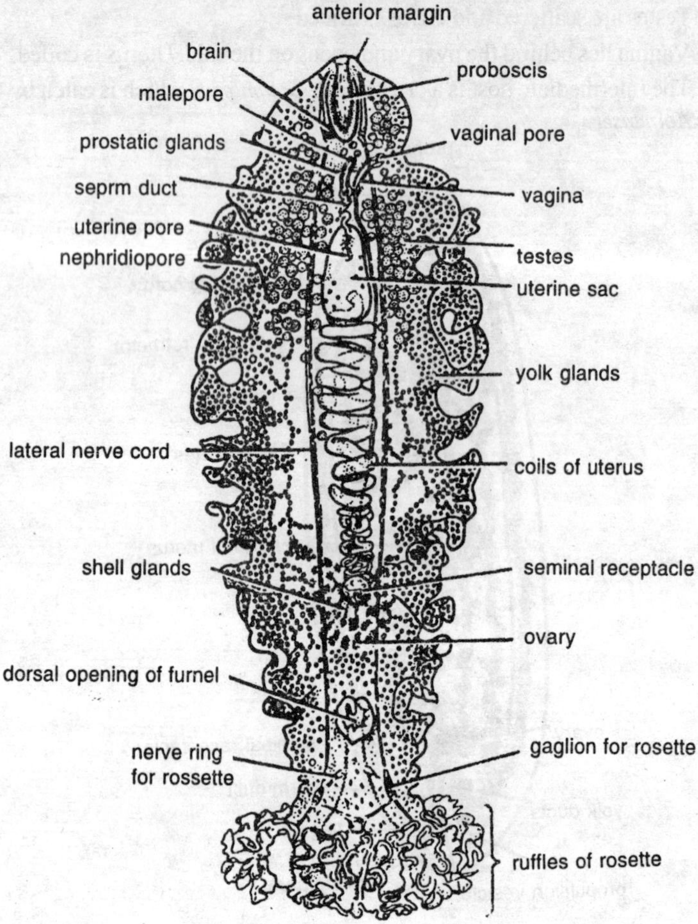

Fig. 1.21. *Gyrocotyle.*

3. Anterior end bears a large opening that leads into a highly muscular protrusible mass, the proboscis.
4. Posterior end bears a rosette-shaped adhesive organ, surrounding a funnel shaped depression.
5. The margins of body are generally ruffled.
6. Bisexual male reproductive system comprise scattered testes, sperm duct and penis papilla. Female reproductive system comprises ovary, oviduct and uterus.
7. Follicles of yolk gland scattered throughout.

TETRARHYNCHUS

Phylum	-	Platyhelminthes
Class	-	Cestoda
Subclass	-	Eucestoda
Order	-	Trypanorhyncha
Genus	-	*Tetrarhynchus*

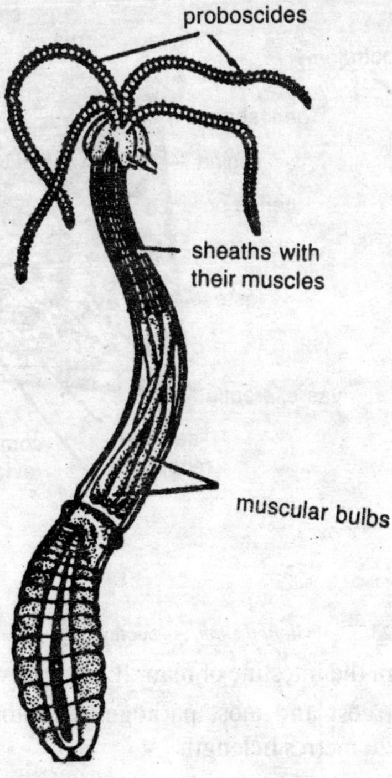

Fig. 1.22. *Tetrarhynchus.*

1. It is an endoparasite in the intestine of elasmobranch fishes.
2. Body is long divisible into scolex and proglottids.
3. Scolex is long and differentiated into a long proximal part containing the proboscis apparatus and a distal part bearing the bothria or sucker.
4. Scolex bears four suckers each with a eversible proboscis armed with spines. The proboscis are retractile in muscular sheaths which end in muscular bulbs.
5. Each proglottid contains a complete set of reproductive organs.

DIPHYLLOBOTHRIUM (DIBOTHRIOCEPHALUS)

Phylum	-	Platyhelminthes
Class	-	Cestoda
Subclass	-	Eucestoda
Order	-	Pseudophyllidea
Genus	-	*Diphyllobothrium*
Species	-	*latum*

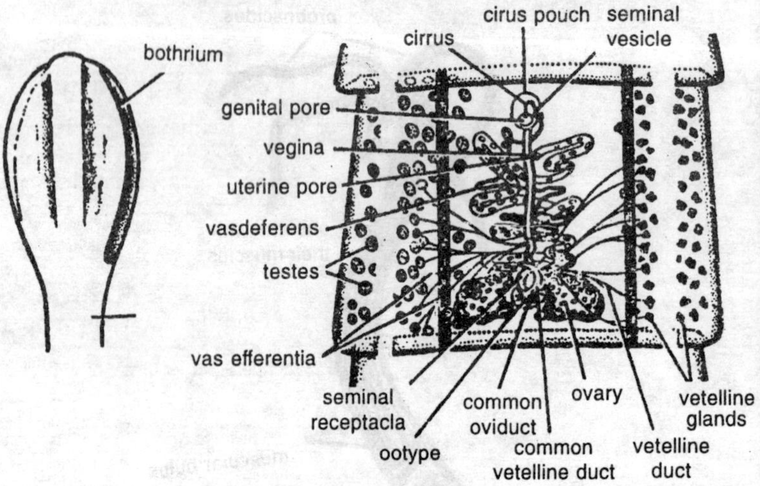

Fig. 1.23. *Diphyllobothrium. A- Scolex; B-A mature proglottid.*

1. It is found in the intestine of man. It is world-wide in distribution.
2. It is the largest and most pathogenic cestod of human beings, measuring 20 metres in length.
3. Body is differentiated into scolex. neck and proglottids. The number of proglottids is 3000-4000.

4. Scolex is fusiform and bears two slit-like suckers or bothria.
5. Mature proglottid broader than long.
6. Each proglottid is hermaphrodite and contains complete sets of male and female reproductive organs.
7. Testes are numerous. Vasa efferentia unite to form coiled sperm duct.
8. Ovary is trilobed. Vagina runs posteriorly forming seminal vesicle. Uterus very much twisted.
9. Gonopores ventral.
10. Life-cycle involves two intermediate hosts, one is *Cyclops* (a copipod) and other a fish.
11. It causes bothriocephalus anaemia, erythropenia and haemorrhage.

MONIEZIA

Phylum	-	Platyhelminthes
Class	-	Cestoda
Subclass	-	Eucestoda
Order	-	Taenioidea
Genus	-	*Moniezia*
Species	-	*expansa*

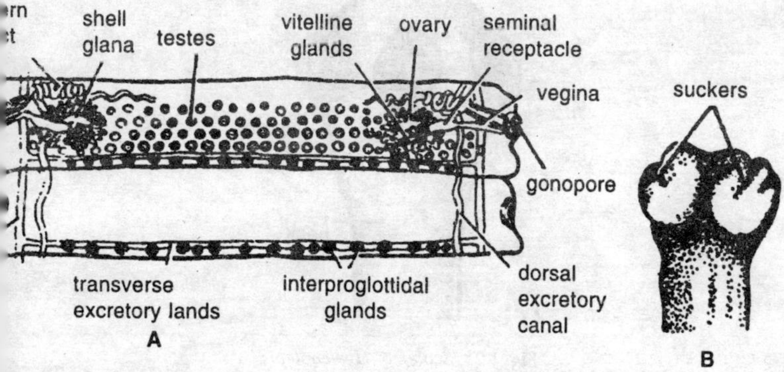

Fig. 1.24. *Moniezia*. A- Two proglottid, B- Scolex.

1. It is found in sheep, cattles and other ruminants.
2. Body is divisible into scolex, neck and proglottids.
3. The scolex is small with prominant suckers. The rostellum and hooks are absent.
4. Each proglottid is broader than long and contains a double set of reproductive organs.

5. Genital ducts are dorsal to osmoregulatory canals.
6. Testes are numerous scattered throughout the centre of proglottid. Cirrus in cirrus pouch.
7. Ovaries are in the form of an open fan. Uteri form networks.
8. On one side vagina is dorsal to cirrus pouch and on other ventral.
9. The uterus develops as a network of canals, but latter breaks up into a large number of uterine capsules, each enclosing 3-30 embryos. Intermediate host is non pathogenic mite, *Galumna.*
10. It is not so pathogenic but sometimes may cause diarrhoea or anaemia.

HYMENOLEPIS NANA

Phylum	-	Platyhelminthes
Class	-	Cestoda
Subclass	-	Eucestoda
Order	-	Taenoidea
Genus	-	*Hymenolepis*
Species	-	*nana*

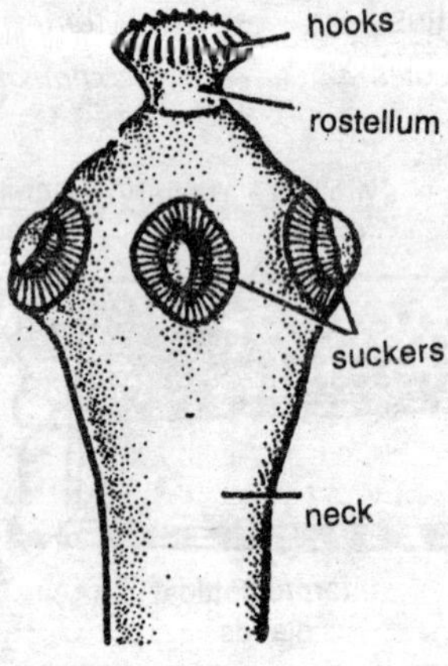

Fig. 1.25. *Scolex of Hymenolepis.*

1. It is commonly called the dwarf tapeworm of man. It lives in the intestine.
2. It is world-wide in distribution.
3. It measures 25-40 mm in lengths and has about 100-200 proglottids.
4. The rostellum is well-developed and retractile. It bears a single circlets of 20-30 hooks.
5. Neck is long and slender.

6. There are three testes with varying arrangement. Cirrus pouch is large with external and internal seminal vesicle and with an accessory sac.
7. Life-history is simple without any intermediate host.
8. The inner membrane of the hexacanth bears long, wavy filaments.
9. Larvae develop in the intestinal villi while the adult lives in the intestine.
10. It causes abdominal pain and diarrhoea.

DIPYLIDIUM CANINUM

Phylum	-	Platyhelminthes
Class	-	Cestoda
Subclass	-	Eucestoda
Order	-	Taenioidea
Genus	-	*Dipylidium*
Species	-	*caninum*

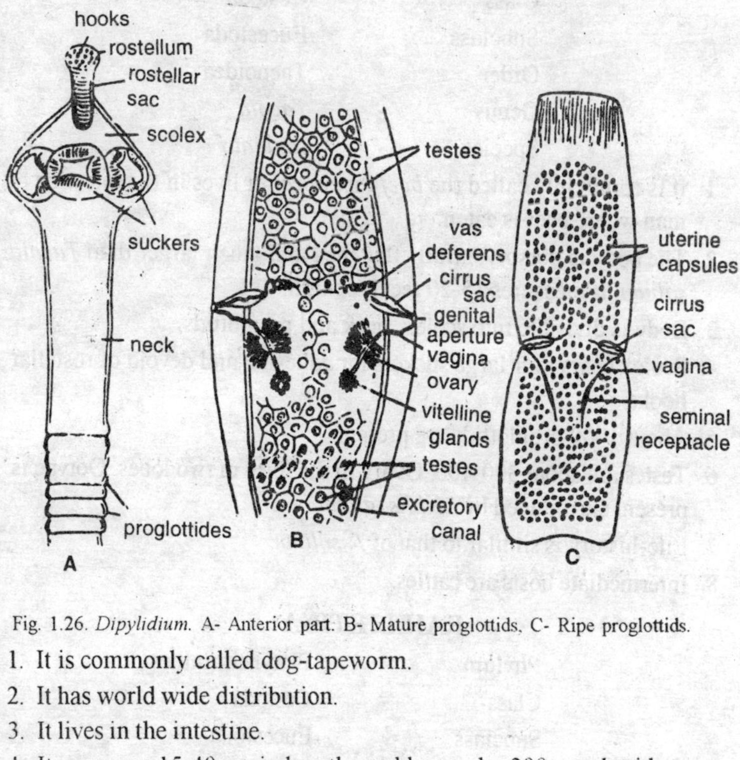

Fig. 1.26. *Dipylidium*. A- Anterior part. B- Mature proglottids. C- Ripe proglottids.

1. It is commonly called dog-tapeworm.
2. It has world wide distribution.
3. It lives in the intestine.
4. It measures 15-40 cm in lengths and has under 200 proglottids.

5. The scolex is somewhat triangular and bears four deeply cupped suckers.
6. Rostellum is rectractile into a rostellar sac in scolex. It bears about 60 hooks in 3-7 rings.
7. Mature proglottids are long and barrel shaped having a double set of male and female sex organs. The genital pores at the middle of both lateral margins.
8. Numerous testes are present before or behind the female genitalia.
9. The uterus first develops as a network of canales, but latter breaks up into a large number of uterine capsules, each enclosing 3-30 embryos.
10. Intermediate host is dog-fleas, *Ctenocephalus canis* and *Pulex irritans.*

TAENIA SAGINATA

Phylum	-	Platyhelminthes
Class	-	Cestoda
Subclass	-	Eucestoda
Order	-	Taenoidea
Genus	-	*Taenia*
Species	-	*saginata*

1. It is commonly called the *beef tapeworm.* It lives in the intestine of man where beef is eaten.
2. The body is dorso-ventrally flattened and much larger than *Taenia solium*. It measures 15-20 feet in length.
3. Body is divisible into scolex, neck and proglottids.
4. Scolex bears four large suckers for adhesion and devoid of rostellar hooks.
5. About 1000 proglottids are present.
6. Testes are about 300-400. Ovary in the form of two lobes. Ootype is present surrounded by *Mehlis' gland.*
7. Life-history is similar to that of *T. solium.*
8. Intermediate hosts are cattles.

RAILLIETINA

Phylum	-	Platyhelminthes
Class	-	Cestoda
Subclass	-	Eucestoda
Order	-	Taenoidea

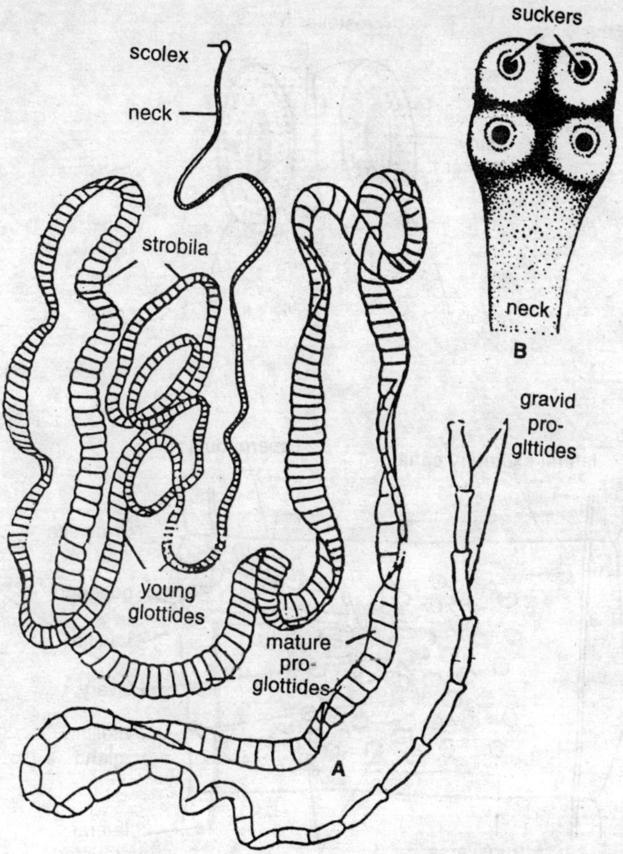

Fig. 1.27. *Taenia saginata.*

Family	-	Davaineidae
Genus	-	*Raillietina*

1. It is commonly found in birds like pigeon, fowl, duck, crow, sparrow etc.
2. It measures 24-39 cm. in length.
3. Scolex has four closely set suckess which are spinose.
4. The cushion-shaped rostellum contains 20 to 110 small hammer-shaped hooks in double-rows.
5. Neck is absent. About 500-700 proglotids are present. having a single set of hermaphrodite reproductive organs.

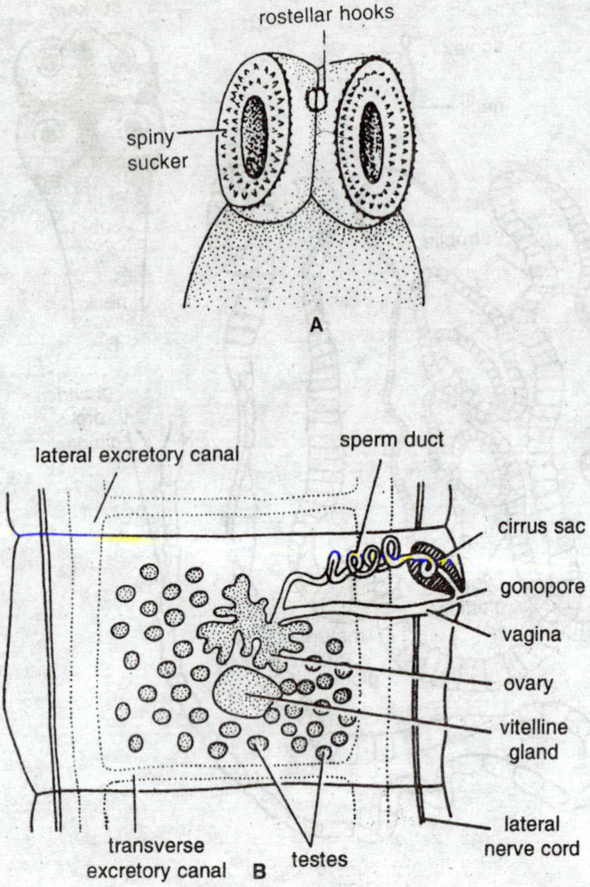

Fig. 1.28.*Raillietina.* A- Scolex, B- Mature proglottid.

6. Male reproductive organs consist of about 20 lobes of testes, a short coiled sperm duct, and a cirris sac with cirrus.
7. The female reproductive organs consist of some what median placed ovary, vitelline gland, seminal recapticle (present in vagina) and uterus.
8. The uterus contains masses of tubules or capsules with eggs.
9. Genital pores alternate irregularly.
10. The animal is digenetic. The primary host is a bird and intermediate hosts are house flies and ants, in which a cysticercoid larva develops.

COTUGNIA

Phylum	-	Platyhelminthes
Class	-	Cestoda
Subclass	-	Eucestoda
Order	-	Taenoidea
Family	-	Davaineidae
Genus	-	*Cotugnia*

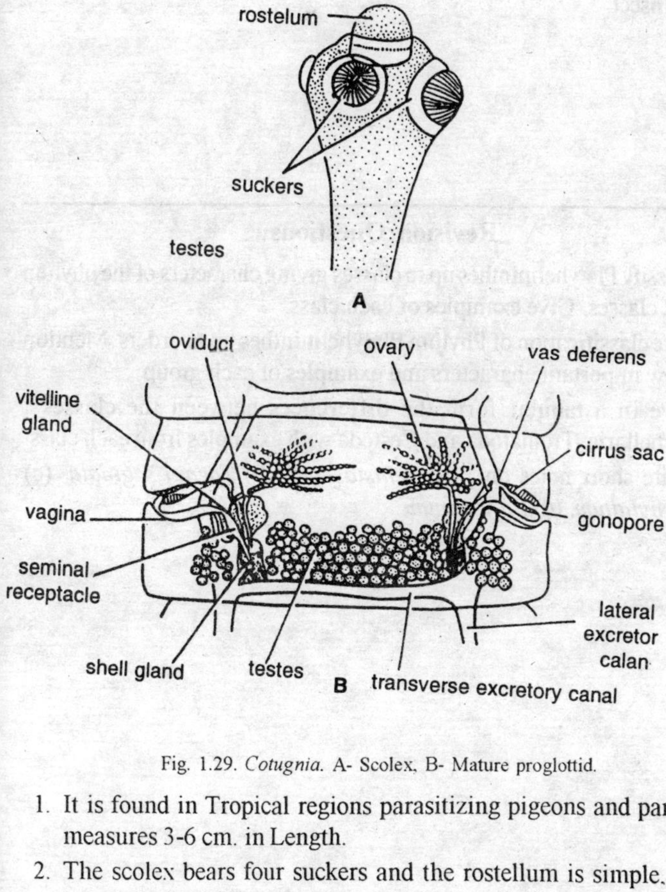

Fig. 1.29. *Cotugnia.* A- Scolex, B- Mature proglottid.

1. It is found in Tropical regions parasitizing pigeons and parrots. It measures 3-6 cm. in Length.
2. The scolex bears four suckers and the rostellum is simple, with a large number of hammer-shaped hooks arranged in two rows.

3. Proglottids are very short, broader than long, except the last ones which are longer than broad.
4. Proglottids bear double set of hermaphrodite reproductive organs.
5. The longitudinal muscles alternate with transverse fibres.
6. Male reproductive organs contain several testes scattered in the posterior region of proglottid, a coiled vas deferens and a cirrus sac enclosing a cirrus on each side.
7. Female reproductive organs contain an ovary, an oviduct and seminal vesicle on each side.
8. The cysticercoid stage passes in an intermediate host which is usually an insect.

Revision Questions

1. Classify Platyhelminthes up to classes giving characters of the phylum and classes. Give examples of each class.
2. Give classification of Phylum Platyhelminthes up to orders. Mention a few important characters and examples of each group.
3. Give in a tabular form the differences between the classes : Turbellaria, Trematoda and Cestoda with examples from each class.
4. Write short notes on : (a) *Schistosoma,* (b) *Taenia saginata,* (c) *Diphylidium,* (d) *Polystomum.*

2

Dugesia (Planaria)

Planaria or *Dugesia* belongs to class Turbellaria. The Turbellarians are mostly free-living, unsegmented flatworms that are clothed with a cellular epidermis, which is usually ciliated, sometimes only in parts. Turbellariens are primarily aquatic, and the great majority are marine. Although there are a few pelagic species, most of them are bottom dwellers that live in mud or sand, under stones and shells, or on sea weeds, *Dugesia* is among them. The following description mostly belongs to the common fresh water genus, the *Dugesia* formarly called *Planaria*. It is a well-known representative of the Turbellaria.

SYSTEMATIC POSTION

Phylum	-	Platyhelminthes
Class	-	Turbellaria
Order	-	Tricladida
Family	-	Planariidae
Genus	-	*Dugesia*

Habits and habitat

Dugesia is commonest free-living Platyhelminthes. They are usually found on the vegetation or under the stones in ponds, lakes, marshes, rivers and small streams. They generally inhabit pure, cool and clear water. Some species remain in currents, while others like to remain in quiet water. When active, they glide over the surface but when at rest they crawl down to the bottom or under other smooth objects, where light is not bright. *Dugesia* is found under the stone generally attached in great numbers, if pieces of meat are dropped in water near the bank, they loosen their hold.

If starved they become smaller, but if well-fed they lay cocoons containing eggs which hatch to give young.They are carnivorous and extremely cannibalistic.

MORPHOLOGY

Shape, Size and Colour

These are small, thin, dorseventrally flattened, bilaterally symmetrical animals measuring about 12 mm long. Colouration is due to the dissolved pigments present beneath the epidermis. The colour may be dark yellow or olive, brown or blackish-brown and may have stripes of white, orange or yellow colour. The colouration is light or whitish on ventral side.

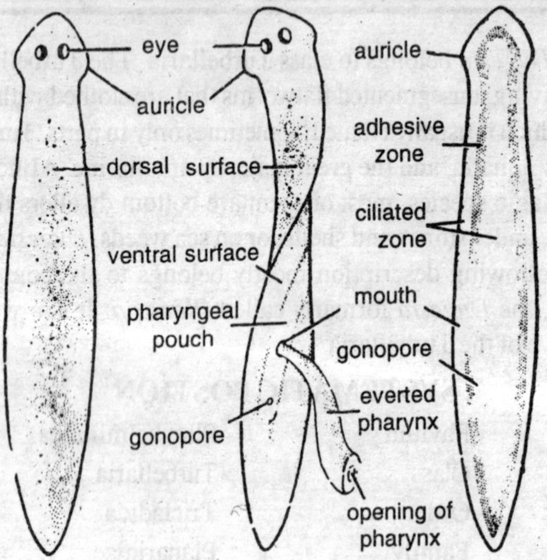

Fig. 2.1. *Dugesia*. External features. A- Dorsal view, B- Lateral view, C- Ventral view

Structure

There are two ends in the body. Anterior end is blunt, while the posterior end is tapering. Anterior end is almost square or triangular with its apex pointing forward. This portion of the body is known as *head*, because with it animal moves forward. There are two lateral projections, called *auricles* or *ears* at the base of the head. But these ears are not sensory organs. On the dorsal side of the head, near the anterior end, there are two black *eye-spots*. Behind the auricles the body is somewhat constricted to form a so called *neck*. On the ventral side of body, near its middle, is found the mouth through which *proboscis* or *pharynx* protrudes. Thus, in *Dugesia* mouth is not situated on the head. The muscular walls of the pharynx serve for capturing the food. Behind the mouth in mature worms is situated a small *genital pore*. There are also excretory openings.

the *nephridiopores,* on the dorsal surface situated laterally, but they are very minute and are, therefore, invisible. The ventral surface of the body is covered with cilia, which help in locomotion.

Body wall

The body of a planarian is covered by a ciliated epidermis. The epidermal cell on its free surface also show a number of microvilli. Cilia are more prominent on the ventral surface. The epidermal cells rest on a basement membrane. Among the epidermal cells the following other types of cells are also present:

(i) *Gland cells* secreting adhesive substances.
(ii) *Rhabditogen cells* which secrete and contains rhabdites.
(iii) *Sensory cells.*

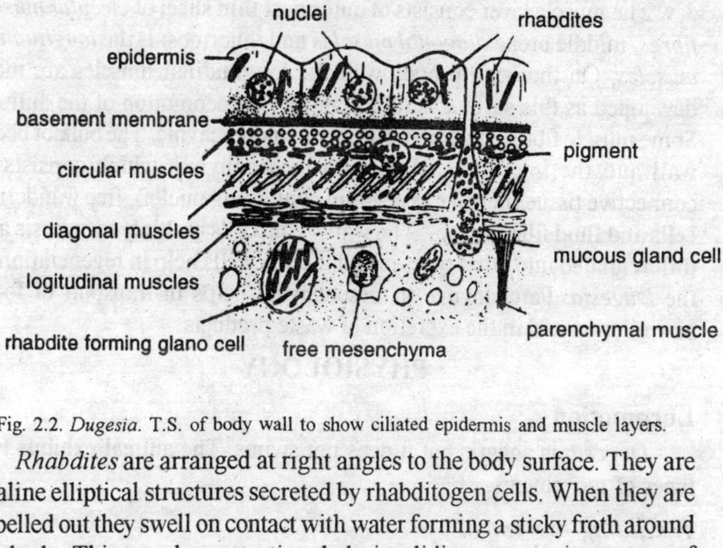

Fig. 2.2. *Dugesia.* T.S. of body wall to show ciliated epidermis and muscle layers.

Rhabdites are arranged at right angles to the body surface. They are hyaline elliptical structures secreted by rhabditogen cells. When they are expelled out they swell on contact with water forming a sticky froth around the body. This may be protective, help in gliding or even in capture of food. Some workers like *Reisinger* and *Kelbetz* (1964) suggest that the rhabdites and nematocysts are evolutionary related.

Below the epidermis are granules and rods of pigments. The gland cells are unicellular, some occur in the epidermis but most of them are in the mesenchyme. They have long necks opening on the surface and secrete mucous.

Below the epidermis is a thin elastic basement membrane at which the epidermal cells rest. It maintains the general form of the body and provides surface for the attachment of underlying muscles.

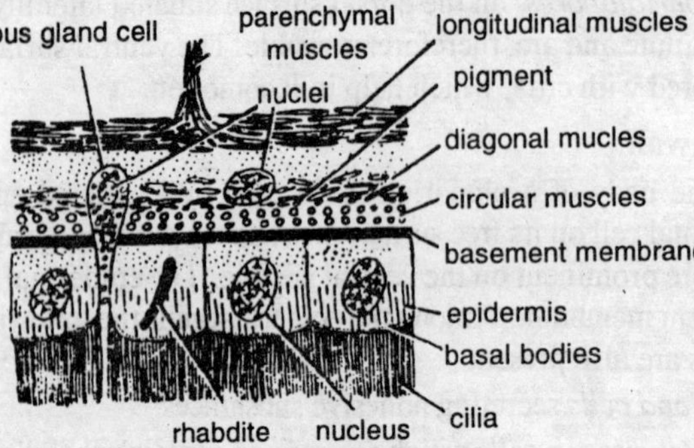

Fig. 2.3. *Dugesia*. V.L.S. Ventral body wall.

The muscle layer consists of outermost thin sheet of *circular muscle fibres*, middle broad *diagonal muscles* and innermost is the *longitudinal muscles*. On the ventral body wall the longitudinal muscles are more developed as this surface is important in the locomotion of the animal. Some muscle fibres run dorsoventrally in mesenchyme. The bulk of body-wall and the interior is filled with mesenchyme, which consists of connective tissue (net-like syncytium containing nuclei), free wandering cells and fluid filled spaces. The wandering cells multiply by mitosis and differentiated into other types of cells. These cells help in regeneration in the *Dugesia*. Parenchyma or mesenchyma helps in transport of food, assimilation and in the excretion of waste products.

PHYSIOLOGY

Locomotion

Dugesia is aquatic but it does not swims. The animal exhibits two types of movements.

1. Gliding movement

In gliding movement cilia present on the ventral surface beat against a track of mucus secreted by the mucous glands. As a result of ciliary action, the animal is propelled forward in a gliding manner. The wave of ciliary action proceeds from the anterior to the posterior end and the head of the animal is slightly elevated.

2. Crawling movement

It is brought about by the contraction of muscles.The wave of contraction extends from anterior to the posterior end, contraction of circular and dorsoventral muscle elongate the body, the anterior is held

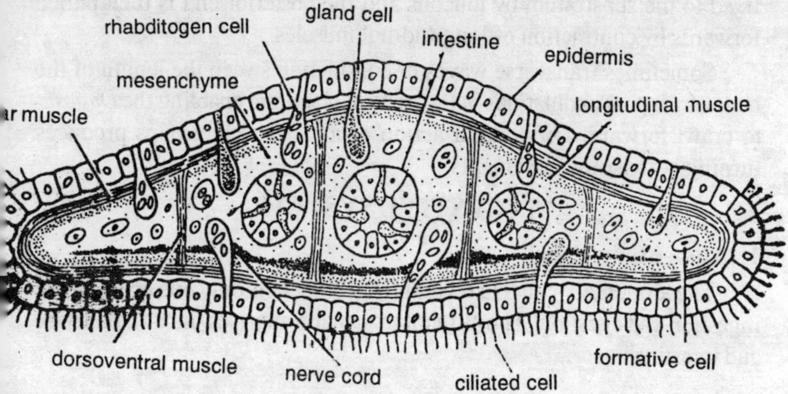

Fig. 2.4. *Dugesia.* T.S. of body showing the disposition of muscles, nerve cord etc.

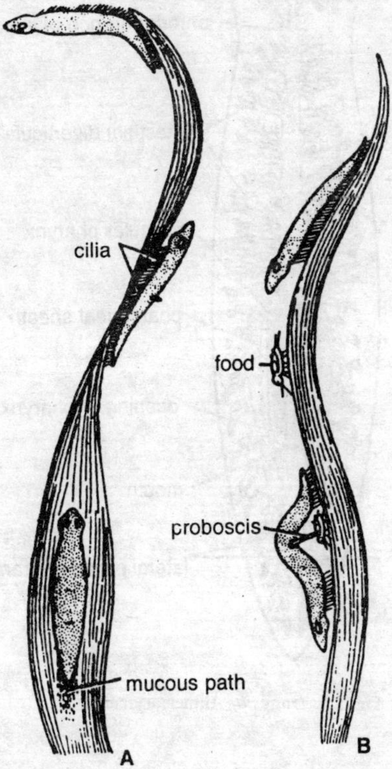

Fig. 2.5. *Dugesia.* A—Locomotion, B—Feeding.

fixed to the substratum by mucous and the posterior end is then pulled forwards by contraction of longitudinal muscles.

Sometimes transverse waves of contraction sweep the length of the body causing its undulations in the vertical plane and enabling the *Dugesia* to crawl forward. Differential action of local muscle groups produces turning or twisting movements.

DIGESTIVE SYSTEM

1. Alimentary canal

The alimentary canal is incomplete i.e. there is only one opening for ingestion and egestion. The alimentary canal comprises: mouth, pharynx and intestine.

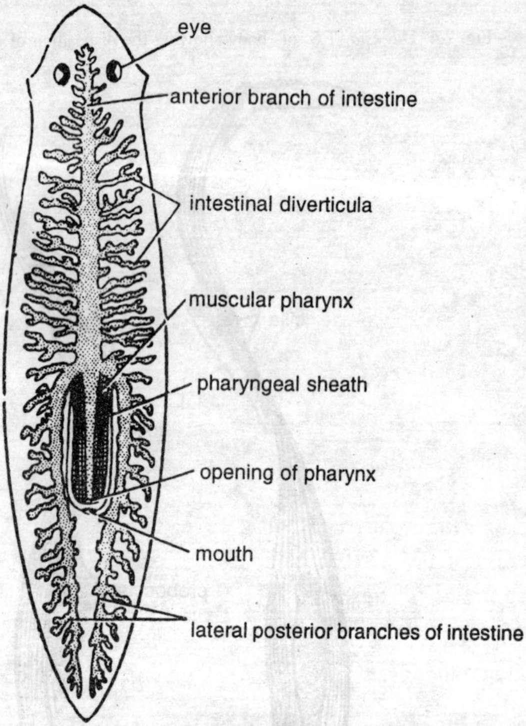

Fig. 2.6. *Dugesia.* Alimentary canal.

Mouth

The mouth is an oval or rounded aperture situated mid-ventrally a little behind the middle of the body. It is surrounded by radial and circular muscle fibres from the subepidermal musculature. The muscle fibres regulate the opening or closing of mouth. It leads into a large cylindrical elongated chamber. The *pharyngeal pouch*.

Pharynx

The pharynx is a tubular organ situated within the pharyngeal pouch projects posteriorly. The pouch is bounded by a muscular sheath, the *pharyngeal sheath*. A space sometimes called *buccal cavity* is present between mouth and the opening of pharynx. The free end of the pharynx projected out of the mouth during feeding. The projecting pharynx is called *proboscis*. This is brought about by the action of muscular wall. Pharynx opens into the *intestine* by way of a short *oesophagus*.

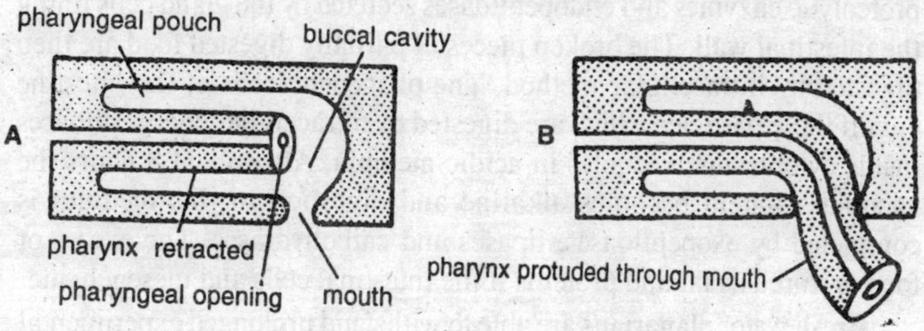

Fig. 2.7. *Dugesia.* Diagrammatric representation of : A- Retracted pharynx, B-Protruded pharynx.

Intestine

The intestine immediatly divide into three branches. The median branch extends forwards upto the head while the other wo lateral branches extend backwards upto the posterior end, one on either side of pharyngeal cavity. All the three branches give off numerous branching and blind diverticulae which end into the mesenchyme. The much branching intestine is a means of increasing the suface aea for digestion, absorption and distribution of food.

Histology of alimentary canal

The pharyngeal wall shows a complicated histological structure. It consists of nine layers from the surface to the lumen : epithelial cells, longitudinal muscle layer, circular muscle layer, outer gland cells, nerve plexuses, inner gland cells, longitudinal muscle layer, circular muscle layer and an endodermal epithelial lining.

The intestine is simple and thin consists of a single-layered epithelium. the gastrodermis. The gastrodermal cells are of two types *phagocytic cells* and *granular cells*.

Food and feeding

Normally *Dugesia* is *holozoic* and canivorous subsisting for the most part on crustaceans, insect larvae and small worms. During feeding, it glides through the action of cilia and the adhesive glands secrete adhesive secretion. When prey comes in contact with the secretion it becomes firmly adherent. *Dugesia* then folds the anterior end of its body over the prey and thus immobilizes it. The pharynx is then produced out through the mouth and through its peristaltic action sucks up small particles of the prey into its intestine.

Digestion

Digestion is initially extracellular. Food is broken down by pharyngeal proteolytic enzymes and endopeptidases secreted by the gland cells lining the intestinal wall. The broken pieces of partially digested food are then digested by intracellular method. The phagocytic cells of the intestine engulf these particles which are digested by intracellular endopeptidases inside cytoplasmic vacuoles in acidic medium. After about 8 hours the vacuolar content becomes alkaline and digestion in this medium is completed by exopeptidases, lipases and carbohydrases. The excess of food is stored as fat and proteins it the intestinal cells and mesenchyme.

Fresh water planarians are able to withstand prolonged experimental starvation. In extreme cases, they utilize part of enteron and all the mesenchyme and reproductive system. In fact, the body volume may be reduced to as little as 1/300 of the original.

Egestion

Since there is no anus, the egestion of undigested food material takes place through mouth. It is brought about by the contraction of the muscles of body-wall.

Respiration

Mode of respiration is aerobic but there are no special organs for this. Oxygen from the surrounding water is directly into the body and carbon dioxide is diffused out through the entire body surface. Flattened body is an admirable adaptation for this type of respiration.

EXCRETORY SYSTEM

The *excretory system* or *protonephridial system* of *Dugesia* consists of a pair of *longitudinal excretory canals* and numerous tiny enlargements known as *flame cells.*

(i) Excretory canals

A pair of longitudinal excretory canals running on each side of the body throughout the length. They are highly inter-coiled and open to the dorsal surface by several minute pores called *nephridiopores.* Anteriorly two canals are united by a *transverse vessel,* in front of eyes. Each canal divides into a number of branches which further divides into very fine blind capillaries which bear flagella at the inner end (*Wilson* and *Webster*, 1974). The terminal structure is composed of *flame cells* or *cryptocytes.*

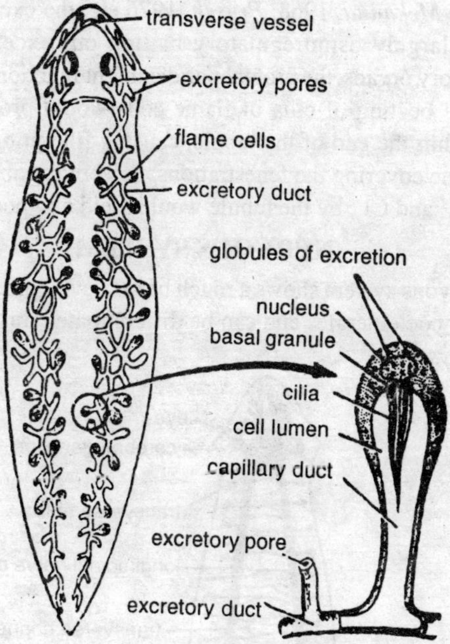

Fig. 2.8. *Dugesia*. Excretory system.

(ii) Flame cells

Each flame cell represents the unit of excretory system. It is a large, tubular, protoplasmic body. It is produced into numerous protoplasmic processes reaching the mesenchyme. Its terminal part is swollen into a bulb and contains a conspicuous nucleus. The flame cell has an intracellular space which is continued into capillaries. The space encloses a bunch of cilia which vibrate giving the appearance of flickering candle flame.

Besides the flame cells, certain glandular cells called the *athrocytes* or *paranephrocytes* occur in close contact with the excretory canals and also help in removal of excretory products.

Physiology of excretion

Excretory substance is collected from the mesenchyme and is transferred into the, cavities of flame cells. The beating of cilia of flame cells causes hydrostatic pressure by which the fluid waste passes into the longitudinal trunks and goes out of nephridiopores.

According to *Storer, Woleott, Guyer, Hickman* etc. the flame cells simply help in excretion of nitrogenous wastes. But according to some workers like *Mekanna*, 1968; *Prusch*, 1976 etc. the excretory system in *Dugesia* is largely osmoregulatory bailing out excess of water. As osmoregulatory organs, the protonephridia might function in the following manner. The beating of cilia of flame cells would produce a negative pressure within the end of the tubule, causing filteration of fluid across the membrane covering the fenestrations. Selective reabsorption of ions, especially K^+ and Cl^-, by the tubule would yield a hyposmotic fluid.

NERVOUS SYSTEM

The nervous system shows a much better development than what we found in the coelenterates and can be differentiated into the *central* and *peripheral systems*.

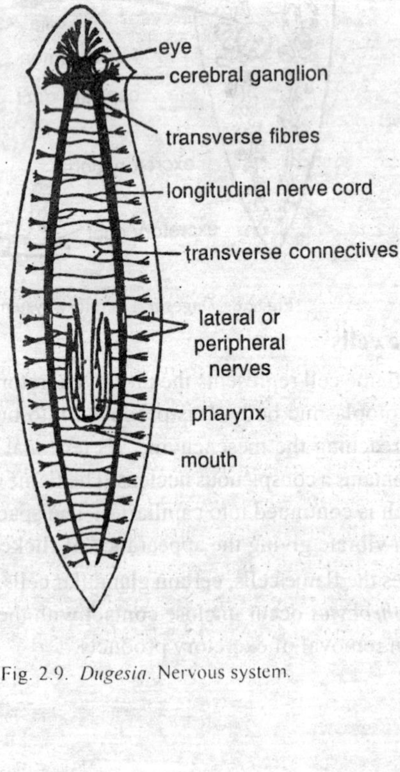

Fig. 2.9. *Dugesia*. Nervous system.

Just below the epidermis and behind the eyespots, there is a bilobed *brain* or *cerebral ganglia* in the form of an inverted 'V'. Both the ganglia are connected by several transverse fibres. From the brain, several thin nerves are given off anteriorly and to the eyes. From the brain two *lateral longitudinal nerve cords* extend posteriorly along the ventral side. The two nerve cords are connected together by *transverse connectives*.

The peripheral system is present in the form of a *sub-epidermal plexus* (just below the epidermis) and a *sub-muscular plexus* below the muscle layers in the mesenchyme.

The nerve-cells or neurons may be unipolar, bipolar or multipolar. All the neurons possessing neurites are similar, These neurites cannot be differentiated into axons and dendrons.

SENSE ORGANS

Dugesia possesses special sensory organs, called *receptors*. These are photoreceptors or eyes, auricular organs, tangoreceptors and rheoreceptors.

(i) Photoreceptors or eyes

The eyes are two dark spots on the dorsal surface of the anterior end, one on either side of the median line. Each eye has the form of a cup whose opening face anterio-posteriorly.

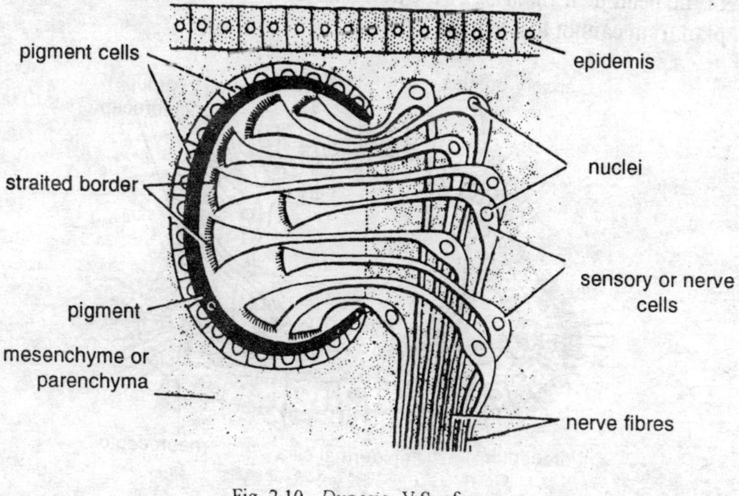

Fig. 2.10. *Dugesia*. V.S. of eye.

Structure of each *eye* is very simple. It is made up of highly pigmented *retina*. Inside the cup are situated numerous nerve-cells, the *fibrillae* of which are in contact with the retina and their opposite ends are continued into an optic nerve, which is joined with *brain*. The nerve cells enter the cup through the opening. Ectoderm over the eye is not pigmented. *Lens* is absent in eye.

Eyes are capable of a crude discrimination of the direction of light. The pigment cup serves as a shield and light can enter only through its opening to stimulate the photosensitive expanded ends of retinal cells, thus the animal can detect the direction of light. The animal is negatively phototactic and is most active at night. If the eyes are removed it can still react to light. However, the reaction is much slower. This shows that it possesses some light sensitive cells over the general body surface.

(ii) Ciliated pits

Besides eyes, there are *ciliated pits* which are also sense-organs. They are situated on either side of the head, having special sensory cells with long cilia. These ciliated pits are connected with nervous network. Tentacles, which are found in some turbellaria with long cilia, are also supposed to the organs of chemical sense, which help in finding food.

(iii) Auricular organs

These are sensitive to chemical and are chemoreceptors. These are in the form of a pair of whitish ciliated grooves, lying one on either side of the head near the base of the auricles. If the auricles are removed the planarian cannot locate food.

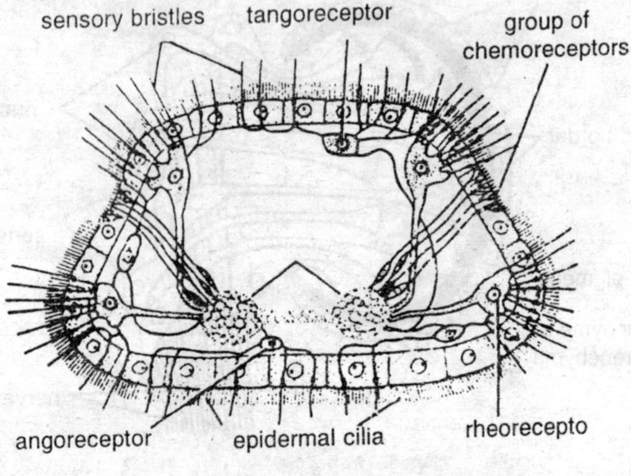

Fig. 2.11. *Dugesia*. Diagrammatic T.S. showing various receptors.

(iv) Tangoreceptors

These are sensitive to touch, temperature etc. A tangoreceptor is a small cell with fine sensory bristles that project over the surface except one that posses to the brain. They are particularly concentrated on the ventral surface around the mouth, auricles and the body margins. Thus ventral surface is positively thigmotactic while the dorsal surface is negatively thigmotactic.

(v) Rheoreceptor

These are sensitive to water current. They are of general occurrence and their sensory processes or bristles project much beyond the level of cilia.

REPRODUCTIVE SYSTEM

Dugesia reproduces asexually as well as sexually.

1. Asexual Reproduction

Asexual reproduction occur by the transverse binary fission. The plane of fission usually forms posterior to the pharynx. The posterior part attaches to substratum by mucous, the anterior part stretches forward until the animal snap into two. Each half regenerates the wanting parts resulting

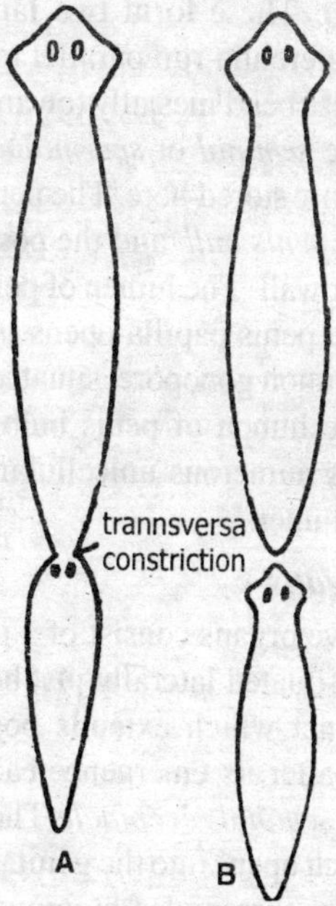

Fig. 2.12. *Dugesia*. Transverse binary fission.

in the formation of two complete planarians. Some planarians fragment into two or more pieces each growing into a complete individual. Asexual reproduction is restricted to some species only. Some species reproduce asexually in one season, and may develop sex organs in the other, but there are some species that reproduce exclusively asexually and rarely develop sex organs.

2. Sexual Reproduction

The reproductive organs are more complicated in these animals. The gonads i.e. testes and ovaries are derived from the mesoderm and there is a system of tubules and chambers in which fertilization occurs.

The animal reproduces in early summer and the gonads are developed temporarily in the breeding season. The *Dugesia* is bisexual or hermaphrodite but self-fertilization does not take place and the cross fertilization is the rule. After breeding season the gonads degenerate and disappear and the worm reproduces asexually.

Male reproductive organs

They consist of a large number of small spherical testes, which are situated on both the sides of the body. Each testis is connected with minute tubes, called *vasa efferntia.* These form two large ducts called *vasa deferentia*. The two vasa deferentia run parallel to one another towards the posterior end, where these bend mesially towards the penis. Each vas deferens dilated to form the *seminal* or *spermiducal vesicle* with thick muscular wall. The sperms are stored here. The penis shows two regions, the anterior large spherical *penis bulb* and the posterior narrow *conical penis papilla* with muscular wall. The lumen of penis is very narrow and called *ejaculatory duct*.The penis papilla opens into the genital atrium which terminates in the common gonopore, situated midventrally behind the opening of mouth. The lumen of penis bulb is formed by *bulbar cavities*. The bulb contains numerous unicellular glands, the *prostate glands* which open into its lumen.

Female reproductive organs

The female reproductive organs consist of a pair of *ovaries,* which are small spherical bodies situated laterally just behind the head. From each ovary arises an oviduct which extends posteriorly through the parenchyma. Immediately after its emergence each oviduct presents a dilation, which is known as *seminal receptacle*.The two oviducts unite to form a common vagina which opens into the genital *atrium*. The oviducts along their whole length are surrounded by numerous *vitelline glands* which open into oviducts by short *vitelline ducts*. Because of the close

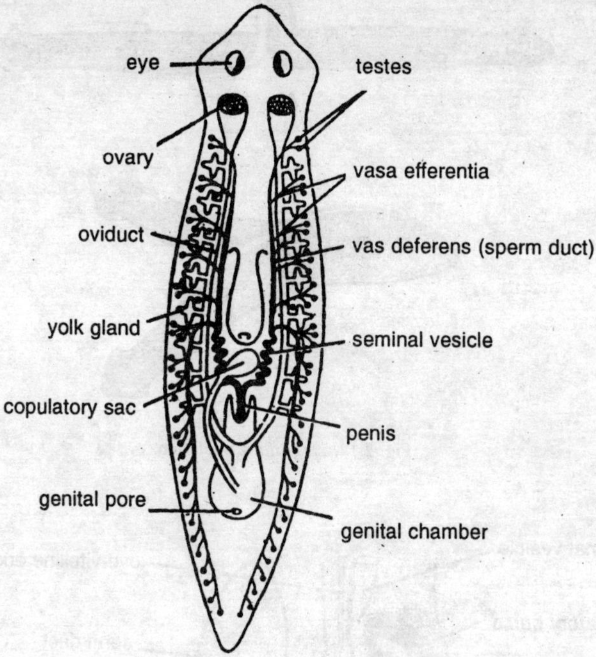

Fig. 2.13. *Dugesia*. Reproductive system.

association of oviduct and vitelline glands, these are sometimes collectively referred to as the *ovovitelline ducts.* Opening into the genital chamber is found a muscular sac-like *copulatory sac* or *uterus* or *bursal canal* The bursal canal opens into a large sac the *bursa copulatrix* which receives penis and temporarily stores the sperms received during copulation. It opens below into female genital atrium.

Copulation

Although *Dugesia* is hermaphrodite, yet *cross-fertilization* is a rule. *Copulation* takes place between the two worms in which their posterior and ventral surface come together and the penis of each is inserted into the *genital atrium* of the other. In this way sperms are deposited in the *seminal receptacles* of the mating partner. The sperms are stored inside the copulatory sac for some time and then travel along the oviducts to reach the seminal receptacles. The eggs are deposited in the seminal receptacles and are fertilized. Thus the fertization is internal. During breeding season, each worm may copulate many times.

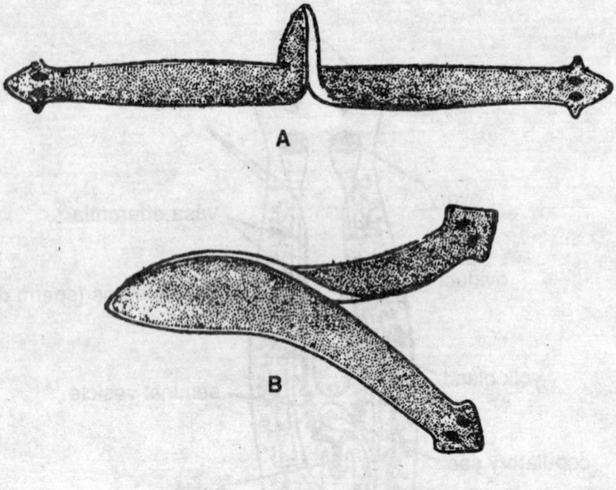

Fig. 2.14. *Dugesia.* Copulation.

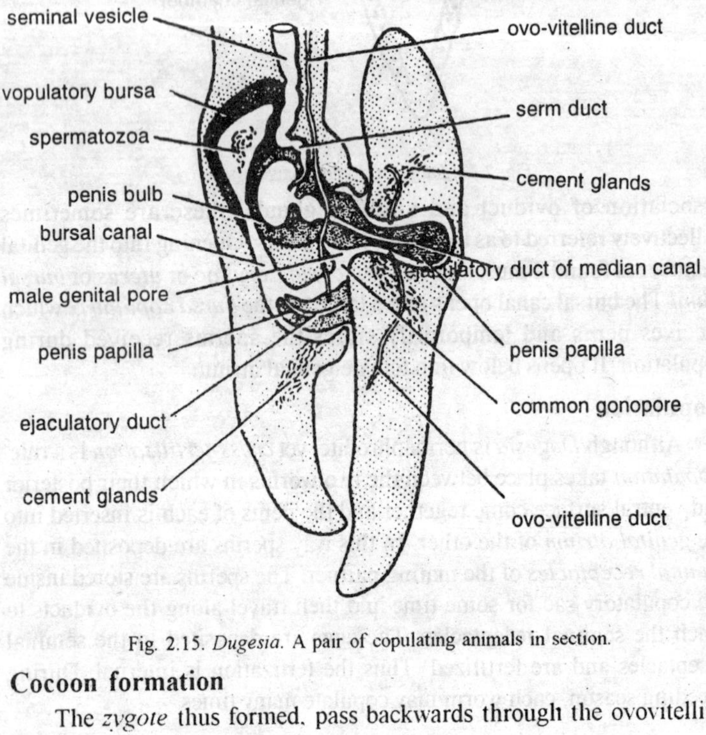

Fig. 2.15. *Dugesia.* A pair of copulating animals in section.

Cocoon formation

The *zygote* thus formed, pass backwards through the ovovitelline ducts and reach the genital atrium of female. In their way they are mingled with yolk cells which are discharged by yolk glands into the oviducts. The diposition of yolk outside the zygote is unique condition in Platyhelminthes. Such eggs are called *ectolecithal* When the eggs are reached in genital atrium 5-10 eggs along with yolk cells are enclosed in a protective, proteinous shell forming a cocoon or an egg-capsule. The capsule is coated by a sticky substance secreted by the cement glands, that surround and open into the common genital atrium. The capsule is now extruded through the genital pore. As it is passing out, sticky coating is drawn out into a sticky stalk that fastens it to some object chiefly the under surface of stones. The capsules are laid in succession at intervals of a few days.

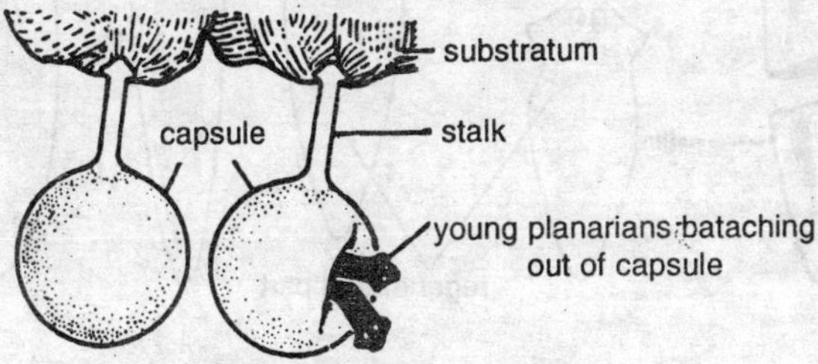

Fig. 2.16. *Dugesia*. Young one hatching from the cocoon

Development

Development starts soon after the cocoon is laid and the hatching occurs in 2 to 3 weeks depending upon the temperature of water. The cleavage is *spiral* and *determinate type*. No larval stage is formed hence the development is direct. A young *Planaria* or *juvenile* glides out of the capsule from each egg.

Regeneration

Like *Hydra, Dugesia* shows extensive regeneration. The lost parts can be easily replaced. If the animal is cut transversely into two halves, the anterior half regenerates a new tail, while the posterior half regenerates a new head. If the animal is cut into three pieces, the middle piece will regenerates head on the anterior side and a tail on posterior side. The regenerating piece retains the original *linear polarity* of the whole animal. Regeneration involves *epimorphosis* and *morphollaxis*. In epimorphosis the missing parts are formed while in morphollaxis, the original parts fit to function with the regenerated parts in the new individuals.

Power of regeneration is maximum near the head and minimum near the posterior end. this is associated with the metabolic rate. The metabolic rate is highest in the head and becomes progressively lower towards the tail. This variation in metabolic rate and regeneration power is called *axial gradient* (*Child*).

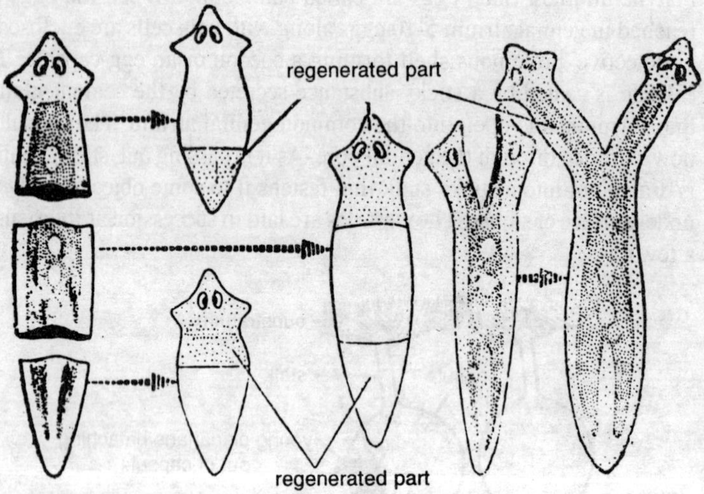

Fig. 2.17. *Dugesia.* Regenaration. A—Transverse, B—Longitudinal.

Regeneration is brought about by the free formative cells or *neoblasts* of the mesenchyme. These cells migrate to the cut surface and by repeated division produce new tissues.

Regenaration is always not perfect. If a sexually mature *Dugesia* is cut transversly between its pharnx and copulatory apparatus, its reproductive organs degenarate and fail to regenarate, but each half grows into asexual animal.

Dugesia shows a *lateral polarity.* If the anterior end is cut longitudinally, each part will grow into a complete head. Thus several headed monster results. The phenomenon is called *heteromorphosis.*

Revision Questions

1. Give an account of the life-history of any flatworm you have studied.
2. Give an account of the reproductive system of *Planaria.*
3. Describe the alimentary canal and modes of feeding and digestion in *Dugesia.*

4. Draw a labelled diagram of the Transverse Section of *Dugesia* passing through pharynx. Discuss the process of locomotion.
5. Discuss the adaptations of *Dugesia* to its environment. Also give a classification of this animal.
6. Give an illustrated account of the reproductive system of any planarian and add a note on its regeneration.
7. Describe the histological structure of body wall of *Dugesia.*
8. Write short notes on : (a) Flame cells, (b) Asexual reproduction in *Degusia,* (c) Eye of *Dugesia,* (d) Rhabdites (e) Parenchyma.

3

Fasciola hepatica

Fasciola belongs to class Trematoda. Most of the trematods living in other animals inhabit the digestive tract or its accessory tubes or cavities, and it seems probable that those species living outside gut are derived from gut dwelling forms. Probably the most important of these flukes, *Fasciola hepatica*, is the first digenetic fluke to have its life-cycle uncovered. *Thomas* (1883) described its life-cycle. *Reinhard* (1957) redescribed its life-cycle in an interesting way.

SYSTEMATIC POSITION

Phylum	-	Platyhelminthes
Class	-	Trematoda
Order	-	Digenea
Family	-	Fasciolidae
Genus	-	*Fasciola*
Species	-	*hepatica*

Geographical Distribution

World wide distribution, occurs in area where sheeps, cattles and goats are raised and there is a niche for limnaied snails. In U.S.A., organisms found in the Gulf Coast, the North-West, Rocky Mountains, Florida etc. In U.S.A., it is endemic.

Habits and Habitat

Fasciola hepatica is derived from Greek and Latin words - (L., *fasciola* = small bandage; Gr., *hepar* = liver) is a *digenetic parasite,* i.e., its life history is completed in two hosts - a primary vertebrate host and a *secondary* or intermediate invertebrate host. Adult *F. hepatica* occurs in the liver and bile passages of the primary host which is usually sheep, but, at times, it may occur in some other vertebrates, like goat, horse, dog, ass, ox, deer, antelope, rabbit, elephant, monkey, etc. Humans are occasional host for *F. hepatica,* and infections are most often obtained through eating watergrass (*Nasturtium officinale*) in salad. A single host may harbour as many as about 200 of these flukes in its liver, which may consequently cease to function normally. This effect is known as *liver-rot.F.hepatica* spends a part of its life history in an intermediate host , a freshwater snail which is either *Limnea truncatula* or some specific species of *Planorbis, Fossaria modicella, Pseudosuccinea* etc.

Shape, Size and Colouration

The body is large, soft and fleshy. It is dorsoventrally flattened and leaf-like having somewhat oval appearance. The size varies from 1.8-5.1 cm in length and about 0.4-1.5 cm. in breadth near the mid body. The colour of body is generally pinkish. The margins are blackish or brownish due to the presence of vitelline glands. The alimentary canal is brownish in colour due to the presence of ingested bile of the host.

Structure

The *anterior end* of *F. hepatica* is triangular while *posterior end* is tapering. The anterior end terminates into a projection, which is called

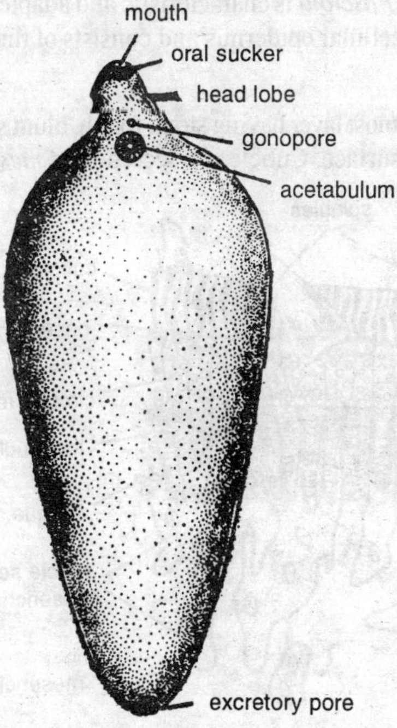

Fig. 3.1. *Fasciola hepatica.* External features.

apical or *cephalic cone* or *head lobe*. At the apex of the head lobe is situated a small opening, the *mouth*, which is surrounded by an anterior muscular sucker, called *oral* or *anterior sucker*. At a short distance behind the head lobe is another saucer-shapted sucker known as *ventral sucker* or *acetabulum,* situated on the ventral side. The ventral sucker is superficial and has no connection with the interior of the body.

Fasciola hepatica was formerly known as *Distomum hepaticum* and the name of the group to which it belongs was known as *Distomidae.* These names were adopted as the mouth and anterior sucker were regarded as two suckers or two mouths (*di*, two;*stoma*,mouth). The word *stoma* was used in trematodes for suckers, which are characteristic structures. But now it is certain that the mouth cannot be claimed as a sucker.

On the dorsal surface there is a single aperture in the middle line at about one-third of the length of the animal from the anterior end. This aperture is the opening of *Laurer's canal*. This aperture appears during breeding season.

Body Wall

Body wall in *Fasciola* is characteristic and adapted to parasitic mode of life. It lacks a cellular epidermis and consists of the following layers:-

1. Cuticle

It is the outermost layer having stout, rough, blunt scales which project over the general surface. Cuticle is a secretion of *mesenchyme,* either of

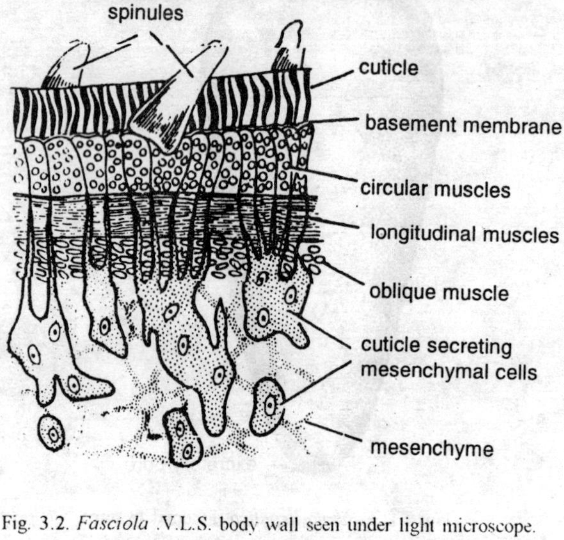

Fig. 3.2. *Fasciola* .V.L.S. body wall seen under light microscope.

special type of cells called *gland cells* formed between muscle-fibres or of general mesenchyme. Cuticle is definitely non-chitinous and is made up of *scleroproteins.* It is situated over the body and protects the liver-fluke from the harmful effects of the different types of juices produced in the host. Below the cuticle is situated a thin transparent layer known as *basement membrane.*

2. Musculature

Immediately beneath the cuticle are three layers of muscles-the outermost *circular* layer, the median *longitudinal* layer and the inner *digonal* layer. In addition to these muscles others make up the suckers and pharynx. Still others, the parenchymal muscles, are dorsoventrally oriented along the lateral regions of the body.

3. Parenchyma or mesenchyma

Immediately benath the musculature are numerous cells, known as *beta cells* or *mesenchymal cells.* Some of the beta cells possess cytoplasmic tubules that penetrate through the muscular layers and reach the cuticle. These specialized beta cells are thought to secrete cuticular materials that reinforce the existing cuticle.

Electron microscopic studies by *Threadgold* (1968) have revealed that the so called *cuticle* of *Fasciola* is actively a thick layer of multi-

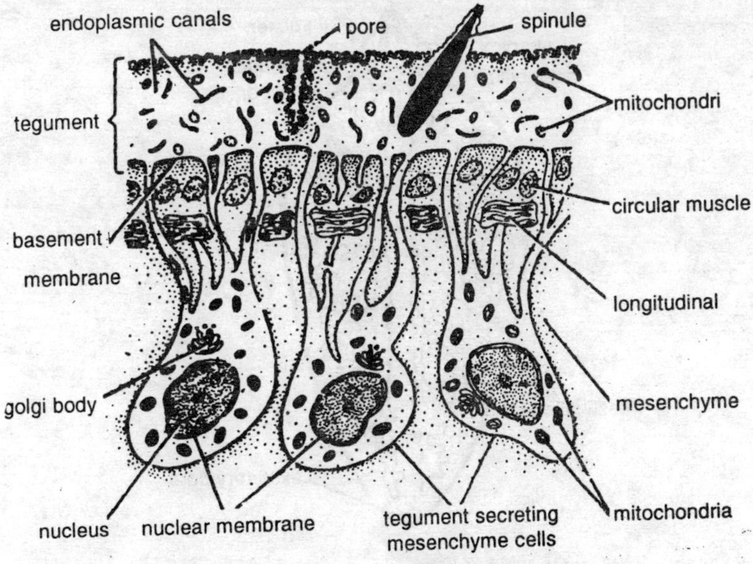

Fig. 3.3. *Fasciola.* V.L.S. body wall seen under electron microscope.

nucleated protoplasm containing mitochondria, canals of endoplasmic reticulum, vacuoles and vesicles. It is termed as *tegument*. It is continuous with *tegument secreting cells* lying embedded in the mesenchyme. The outer surface of tegument is thrown into numerous folds. These increase the surface area considerably for the absorption of fluids of the host. Further, tegument is perforated by fine *pore canals.* Through these canals fluids are absorbed into the mesenchyme.

Functions

The body wall of *Fasciola* is more than an inert protective layer,it is a dynamic structure of great importance in the metabolism of the organism. Although the animals derive a great deal of their nutrients through ingestion, certain chemicals such as certain amino acids, which are necessary as energy sources and as raw materials for the synthesis of complex body chemicals, are absorbed through the body wall. It is also possible that certain metabolic wastes are discharged through cuticle.

DIGESTIVE SYSTEM

The digestive system is complex and highly branched with *mouth* but no *anus.* It consists of the following parts :

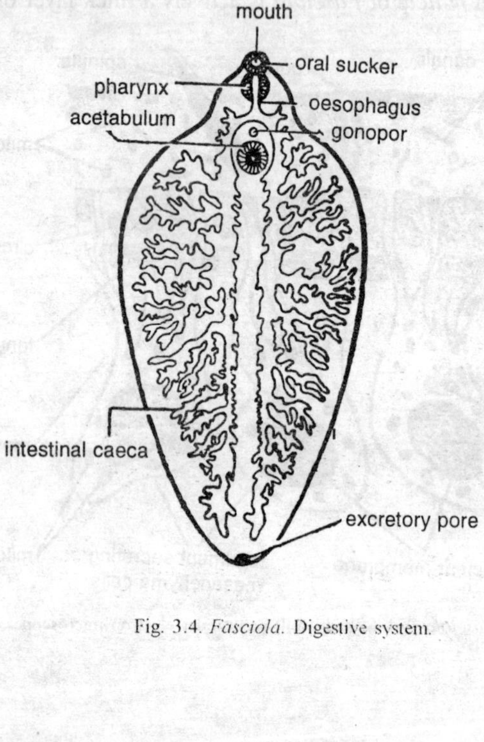

Fig. 3.4. *Fasciola.* Digestive system.

Mouth

It is a small opening situated ventrally at the apex of head lobe and surrounded by oral sucker.

Oral chamber

Mouth opens into a short but funnel-shaped oral chamber.

Pharynx

It leads into a short, rounded, muscular *pharynx* with thick wall and *pharyngeal glands*.

Oesophagus

The pharynx communicates with a short-narrow oesophagus joining the intestine.

Intestine

The intestine immediately bifurcates into two lateral branches, running one on either lateral side of the body upto the posterior end where these end blindly. A large number of caeca or diverticula are given out along either side of its whole length. The diverticula of the outer side are large and highly branched, while those of the inner side are short and unbranched.

Food and feeding

The food of *Fasciola* is bile and blood as it lives in the bile passage of sheep and cattles. It takes about 0.18-0.2 ml of blood per day. Small microscopic debris are also taken as food. The food is taken by the sucking effect of the pharynx and is pumped into the intestine.

Digestion

The exact manner of food digestion is not well understood. Digestion is supposed to be extracellular, carried in the intestinal ceaca. Proteolytic enzymes are reported in the animal. The digested food is diffused into the surrounding mesenchyme. The mesenchyme distributes digested food to every part of body. As earlier mentioned the tegument also helps in the absorption of certain amino acids as well as monosaccharides (glucose and fructose). The reserve food material is in the form of glycogen and fat which are stored in mesenchymal cells and muscles.

Egestion

As there is no anus, the undigested food materials are egested out through mouth.

Respiration

In the *Fasciola*, the respiratory organs are absent so that the respiration takes place through general body surface by the process of diffusion.

Since the animal is an endoparasite, the rate of respiration is very low and is *anaerobic*. In this process, there is no necessity of oxygen. Glycogen undergoes anaerobic glycolysis to form pyruvic acid. Pyruvic acid is decarboxylated to form carbon dioxide and an acetyl group. The latter combines with coenzymes to form acetyl coenzyme A. Acetyl coenzyme A is finally condensed and reduced to form fatty acids. During the process energy is released.

EXCRETORY SYSTEM

The excretory system consists of flame cells or protonephridia connected with an intricate system of excretory ducts.

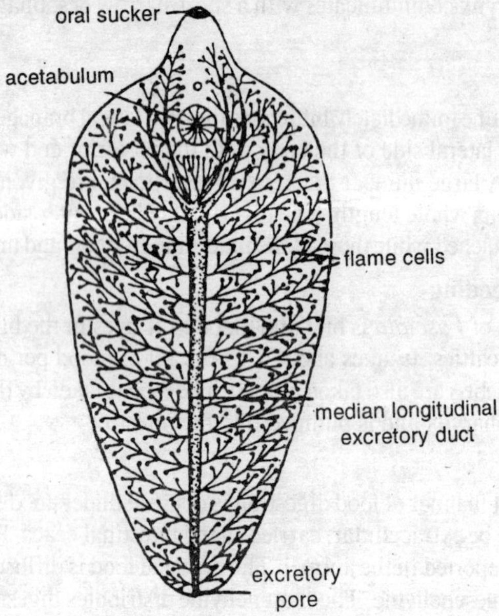

Fig. 3.5. *Fasciola*. Excretory system.

1. Excretory ducts

A single and median longitudinal excretory canal which runs behind the ventral sucker and terminates into the *excretory pore*. It is ramified throughout the body. In the anterior region it receives two ventral and two dorso-lateral excretory canals. The dorsal ducts before opening into the main longitudinal excretory canals are united so as to form a common duct. Main excretory canal as well as dorsal and ventral-lateral excretory canals divide into small capillaries or tubules, each of which ends internally into a *flame cell*.

2. Flame cells

The flame cells are characteristic, each has a thin elastic wall with a nucleus and a cavity containing many long cilia arising from basal granules or blepharoplasts. The cilia in a living animal vibrate like the flickering of the flame, hence named as *flame cells*. The cavity of flame cell is continued with the cavity of capillary. The shape of these cells is irregular due to pseudopodial processes into the mesenchyma.

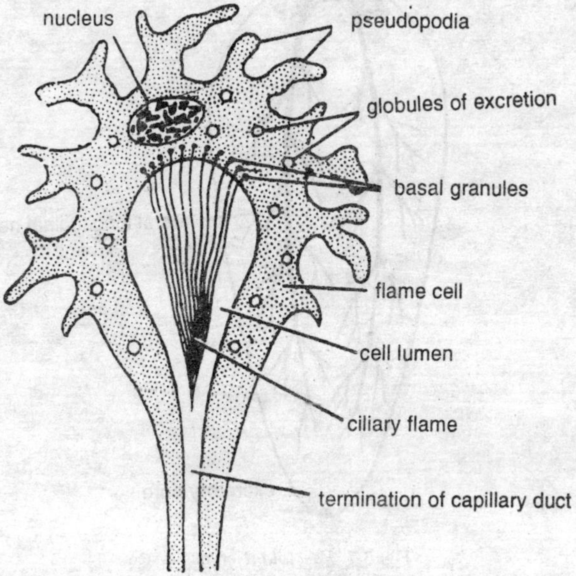

Fig. 3.6. *Fasciola*. A flame cell.

Physiology of excretion

The excretory products are carbon dioxide, fatty acids and ammonia. These are diffused into the flame cells from the surrounding mesenchyme. The fluid moves through the tubule by the action of cilia into the excretory ducts and finally passes out of the body through the excretory pore. All the ducts except the median longitudinal duct are lined with cilia. The course of the flow of excretory products is as follows:

Excretory products → mesenchyme → flame cells
↓
main longitudinal canal ← vessels ← twigs ← capillaries
↓
excretory pore → outside

NERVOUS SYSTEM

The nervous system is well developed in spite of the fact that due to sluggish and parasitic habitats such a large correlation is not need. The nervous system includes:

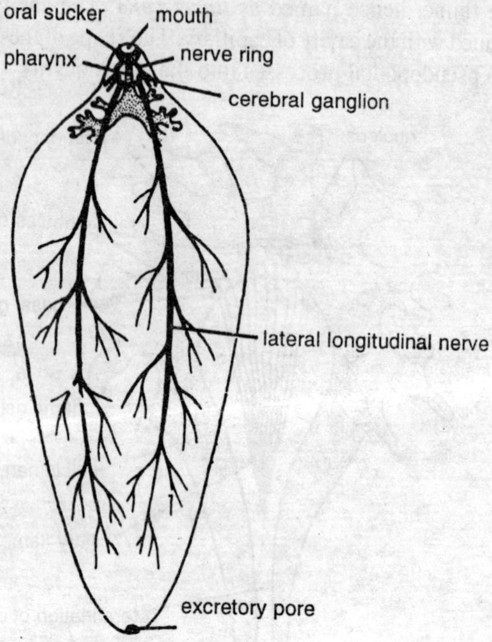

Fig. 3.7. *Fasciola.* Nervous system.

(a) Central nervous system

Two cerebral ganglia are prominant masses of nervous material joined together by a nerve ring around the oesophagus. From each cerebral ganglion arise three longitudinal nerve cords, dorsal, ventral and lateral. Of these the lateral nerves are much longer than the other. They alone reach the posterior end of body. They are interconnected at intervals by transverse commissures.

(b) Peripheral nervous system

From each cerebral ganglion three nerves arise anteriorly and innervate the head lobe and the oral sucker. Fine nerves arise from nerve cords all along their course and supply the various parts of the body.

SENSE ORGANS

Due to parasitic mode of life, special organs are absent. It, however.

possesses *tectile organelles* or *tangoreceptors*. These are bulb-like nerve endings. These are abundant in the sucker region.

Hormones in Fasciola

In *Fasciola* serotonin is present and it seems to play a role in the tissue comparable to that played by adrenalin in mammalian tissue. It is not known whether serotonin acts as a hormonal component of the nervous system, as it is reported to do in some other animals. The tissues of *Fasciola* contain an acetylcholine -like substance and cholinesterase; the specificity of cholinesterase for acetylcholine has not been established, but it has been suggested that these components are related to activity of nervous system.

Insulin has been reported to have an effect on the carbohydrate metabolism but the evidences are lacking.

REPRODUCTIVE SYSTEM

Fasciola hepatica is *hermaphrodite* or *dioecious*. The gonads are well developed and the male and female genital ducts open into a common chamber, the *genital atrium*. It is situated anteriorly in the body and opens to the exterior through the common genital aperture or gonopore, located ventrally in front of the acetabulum.

1. The Male Reproductive Organs

The male reproductive organs are:

(i) ***Testes.*** These are greatly ramified and tubular structures situated in the middle part of the body, one behind the other (*tandem arrangement*).

(ii) ***Vasa deferentia.*** A narrow delicate duct, the *vas deferens*, emerges out from each testis and runs forward to meet the fellow of opposite side forming the common sperm duct just below the ventral sucker.

(iii) ***Seminal vesicle.*** The common sperm duct immediately leads into a large muscular pear-shaped sac, the seminal vesicle, which lies in front of ventral sucker.

(iv) ***Ejaculatory duct.*** The ejaculatory duct arises as fine convoluted tube from the seminal vesicle and runs forward in a zigzag fashion through the cirrus to open into the *genital chamber* through the *male genital pore*.

(v) ***Cirrus and cirrus sac.*** The cirrus or penis is a muscular cylindrical structure traversed internally by the passage of ejaculatory duct. It can be everted out and drawn in through the genital pore and thus helps in copulation. The cirrus and the seminal vesicle both are enclosed in a bag-like *cirrus sac*.

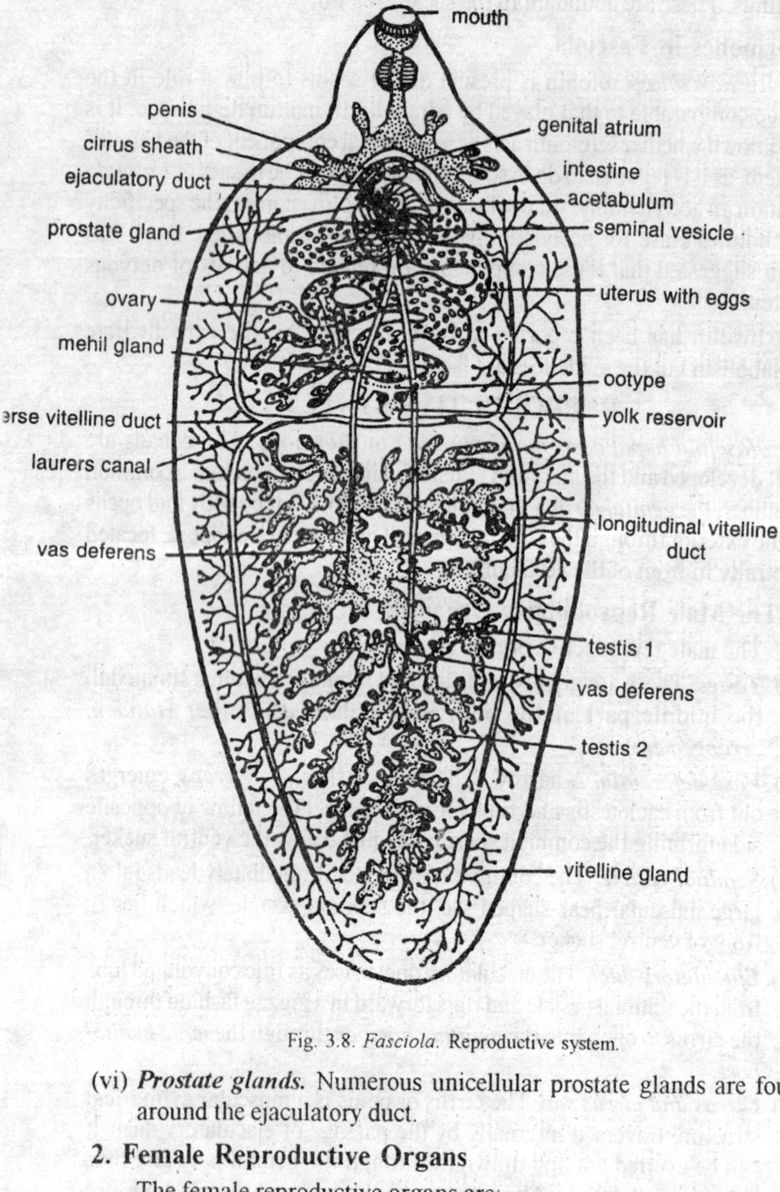

Fig. 3.8. *Fasciola*. Reproductive system.

(vi) ***Prostate glands.*** Numerous unicellular prostate glands are found around the ejaculatory duct.

2. Female Reproductive Organs

The female reproductive organs are:

(i) ***Ovary.*** The ovary is a large, highly branched tubular structure which

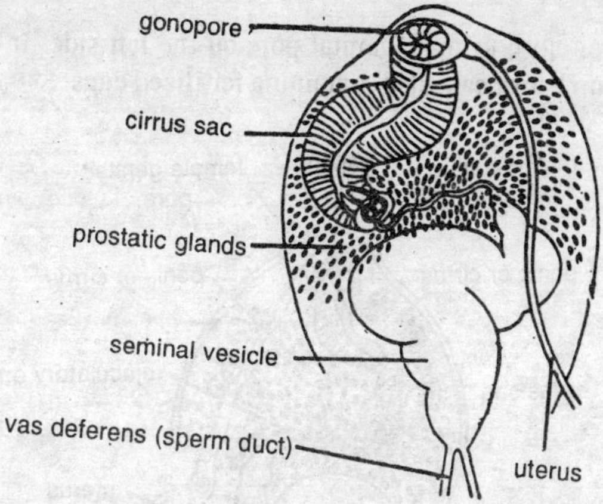

Fig. 3.9. *Fasciola*. Details of male reproductive system.

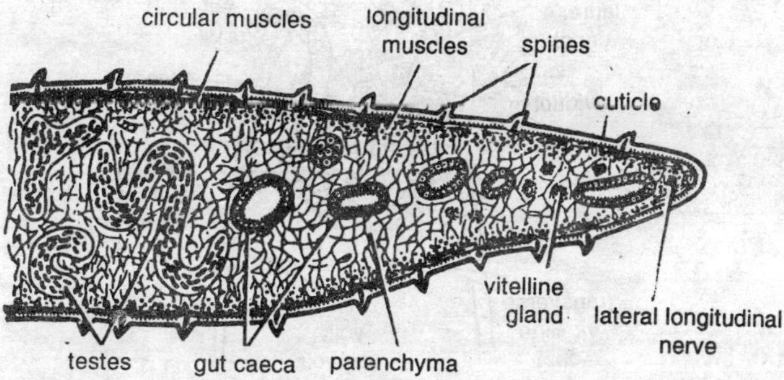

Fig. 3.10. *Fasciola*. Ootype region and egg capsule formation.

is situated on the right side in front of testes in the mesenchyme. It occupies the anterior one-third of the body.

(ii) ***Oviduct.*** From the inner side of ovary arises a short, narrow and convoluted duct, the *oviduct*. It runs down ward to join the uterus.

(iii) ***Uterus.*** It is a long wide and highly convoluted tube that extends upto the genital atrium, opening into it through the female genital aperture, close to male genital pore on the left-side. It contains a large number of capsules containing fertilized eggs.

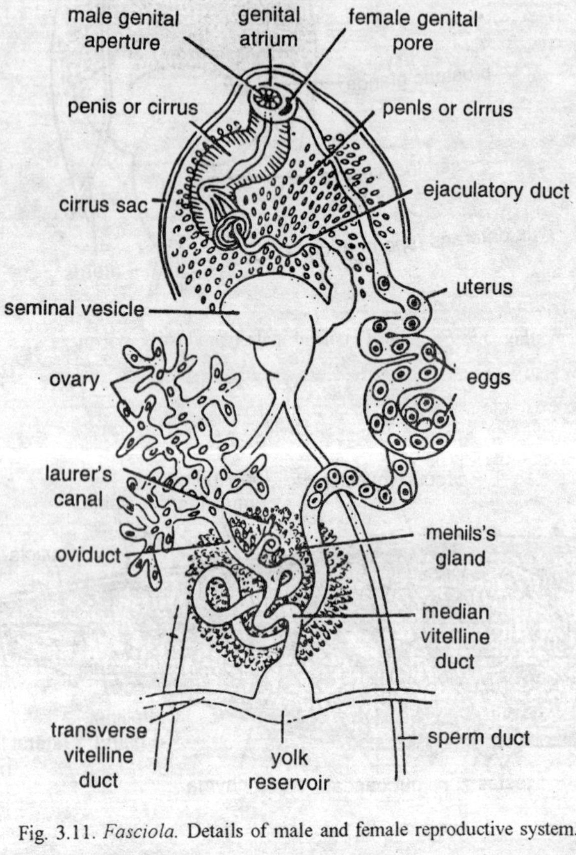

Fig. 3.11. *Fasciola.* Details of male and female reproductive system.

(iv) ***Laurer's canal.*** The Laurer's canal arises from the oviduct and acts as a sperm duct. It develops a temporary opening on the dorsal body surface during breeding season. It may serves as vagina.

(v) ***Mehlis' gland.*** At the junction of oviduct, vitelline duct and uterus a mass of unicellular gland lies known as *Mehlis' glands*. The function of these glands remains unclear. It was believed that these glands were merely responsible for the formation of shell but *Stephenson* (1947), *Rennison* (1953) and *Johri* and *Smyth* (1955) have shown definitely that this cannot be the case, because the shell of eggs is rich in various proteins, phenols and phenolases that are not present in these glands. Several views have been proposed to explain the function of *Mehlis' glands:*

(a) These glands secrete a fluid enchances the hardening or tanning process of newly formed eggs. (b) The secretion of these glands cause release of the shell globules from the vitelline glands. (c) The secretion forms a thin membrane around the cells forming eggs, and the shell globules than built up from within this membrane (d) The secretion lubricates the uterus facilitating passage of the eggs. (e) The secretion activates sperms which are passed down. All or some of these hypothesis may be valid.

(vi) ***Vitelline glands and vitelline ducts.*** The vitelline glands or the yolk glands are in the form of numerous minute vesicles or follicles arranged on either side of the body from one end to the other. A fine duct arises from each vesicle and after uniting with similar ducts from other vesicles opens into the *lateral longitudinal ducts.* The two longitudinal ducts are connected in the middle line by a *transverse vitelline duct.* A median duct arises medially from the transverse vitelline duct and joins the oviduct.

The functions of vitelline glands is now know. In addition to contributing yolk material for incorporation within the egg, the vitelline glands also secrete large globules known as *shell globules* which envelop the developing egg and eventually coalesce and become hardened to form the shell. The hardening process involves the tanning of protein present within the coalesced globules by quinone. Thus vitelline glands contribute both yolk and shell material.

$$\text{Phenol} \xrightarrow{\text{Phenolase}} \text{quinone} \longrightarrow \text{Tanned protein}$$
$$\text{Protein} \longrightarrow \text{Protein} \nearrow$$

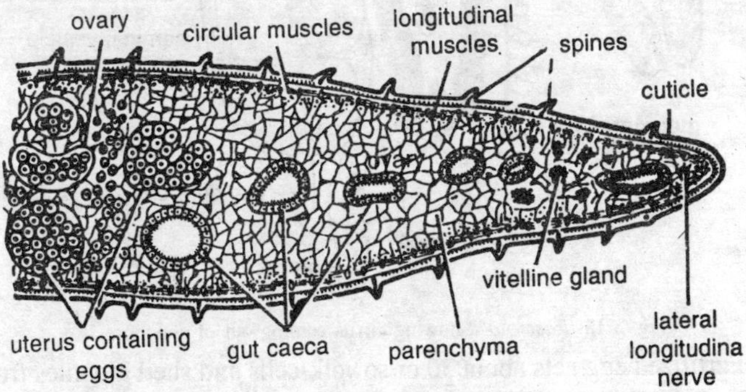

Fig. 3.12. *Fasciola.* T.S. body through ovary and uterus.

Ootype

In some cases the oviduct after receiving *Laurer's canal* and a common vitelline duct swelles to form ootype. In *F. hepatica* according to *Stephenson* (1947) the ootype is absent.

LIFE HISTORY

Copulation and fertilization

Cross-fertilization takes place by insertion of cirrus of one fluke into the *Laurer's canal* of another fluke. Transference of biflagellate sperms directly into the oviduct is possible. Secretion of prostate glands and as mentioned above, the secretion of *Mehlis' gland* keep the sperms active for fertilization. Laurer's canal become invisible after copulation. Fertilization takes place in oviduct. Self-fertilization is also seen here the sperms enter the uterus of same fluke through female genital pore and pass down to reach oviduct where fertilization takes place.

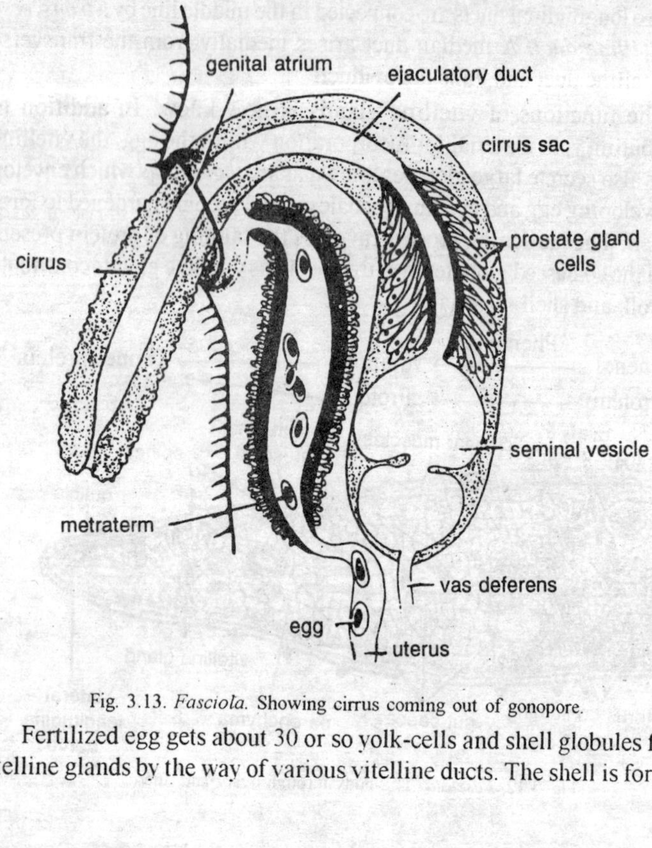

Fig. 3.13. *Fasciola.* Showing cirrus coming out of gonopore.

Fertilized egg gets about 30 or so yolk-cells and shell globules from vitelline glands by the way of various vitelline ducts. The shell is formed with a lid or *operculum* on one side. *Rowen* (1956,57) demonstrated that there is a viscous granular cushion at the opercular end of the egg, situated immediately beneath the operculum. The zygote and a mass of yolk - cells enveloped by the shell constitute the capsule. When complete the *capsule* passes into the uterus.

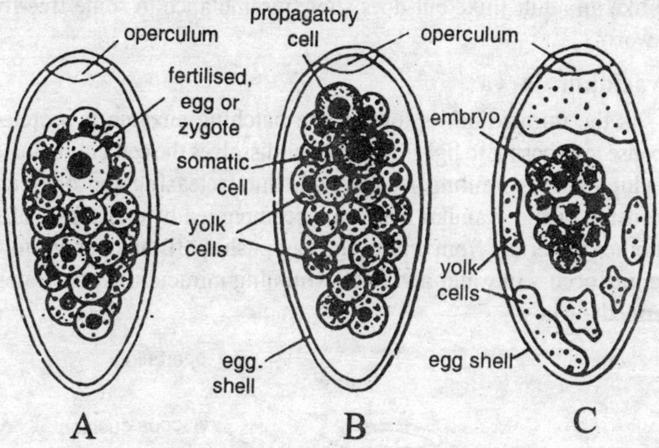

Fig. 3.14. *Fasciola*. Early stages of developments. A—Fertilized egg, B—Two cell-stage, C— Many cell stage.

Capsule formation occurs throughout the year and one fluke may produce about 50,000 capsules. A single fluke may produce a total of 2,00,000 capsules in about 11 years.

The capsules differ in shape and size in different animals. Hence, with the size and shape of capsules the parent flukes are recognized.

Development

Cleavage starts immediately after fertilization. The fertilized egg divides into a small granular *propagative cell* and a large ectodermal *somatic cell*. Somatic cell divides and gives rise to the body-wall of the larva: *Propagative cell* later on divides into a cell similar to *somatic cell*, which takes part in the development of mesoderm, endoderm, and a cell called *germ cell* afterwards reaches the posterior part of the body and forms germ-ball consisting of 14 to 20 germ cells.

Capsules then escape through the gonopore and are liberated into the bile capillaries from where they reach the intestine of the host and are passed out with the faeces.

The development remains arrested while the embryos remain in the faeces. they may survive in wet faecal matter for several months. If washed free, the development of the embryo proceeds. The optimum temperature for development ranges from 10- 30°C. At 30°C the encapsulated embryo differentiates into *miracidium larva* within eight days. The availability of oxygen is also a factor. The fully formed miracidium is a more or less conical little animal covered with ciliated epithelium. It does not look at all alike an adult fluke but does show resemblance to some free-living flat worm.

Miracidium larva

As the miracidium is formed the hatching enzyme is secreted in response to exposure to light. The enzyme dissolves the cementing material by which the operculum is attached, thus releasing the operculum. Expension of the granular cushion, accompanied by exosmosis of salts and other materials from within the egg pushes off the operculum. The hatching occurs in water and free-swimming miracidium penetrates the intermediate host.

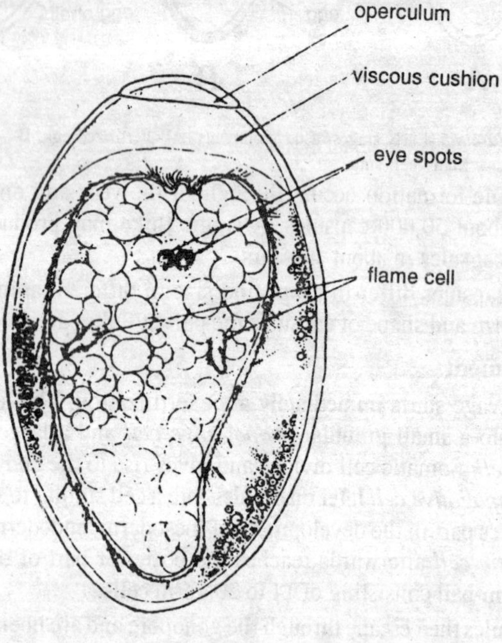

Fig. 3.15. *Fasciola.* Miracidium just before hatching.

Structure of miracidium

It is an elongated, about 0.07 mm long creature with a broad anterior end and a narrow posterior end. Anteriorly there is a *conical* or *apical papilla* or *terebratorium*. It is ciliated except in the region of apical papilla. However, anterior cilia are slightly longer.

The ectoderm is formed by 21 hexagonal plate like cells which are arranged in 5 rows. The I row consists of 6 small plates. The II row contains again 6 plates, III row has 3 plate cells, 4 plates in IV row and 2 plates in V row. The arrangement of the ciliated plates is quite constant and is of considerable value in determining phylogenetic relationship. In *F. hepatica* the cells of first row arranged are two dorsal, two ventral and two lateral in position. The cells of second row are three dorsal and three ventral. In the third row the cells are one dorsal and two venterolateral. The fourth row has two right and two left cells and fifth row has one left and one right cells.

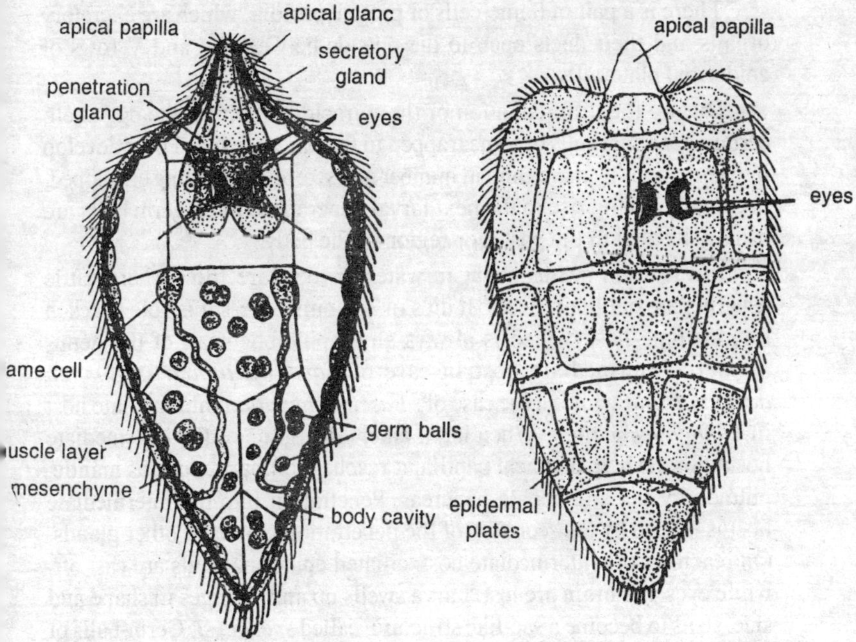

Fig. 3.16. *Fasciola*. Miracidium in ventral and dorsal view.

The ectoderm is followed by musculature consisting of outer circular and inner longitudinal muscle fibres. The musculature is followed by a layer of cells forming the sub-epithelium. The ectoderm, musculature and sub-epithelium forms the body wall of larva.

At the anterior tip of the miracidium is found a mobile apical papilla.

A pair of *cephalic glands* or *penetration glands*, located in the mesenchyme, empty to the outside near the apical papilla. The secretion of these glands aids in dissolving the host's tissue during penetration process. Besides a sac-like multinucleated *apical gland* is located between the two cephelic glands. Previously it was supposed to be the gut of larva.

The sensory organs of miracidium includes two eye-spots. Eyes contain pigments and are situated at the surface of the second row of epidermal plate over the brain. The eyes may be primary receptors involved in the process of hatching. When kept in dark few miracidia hatch, but illumination rapidly induces hatching. The *brain* of the miracidium is a large cephalaic ganglion lying in the parenchyma behind the apical region. It is X-shaped. From this nerve centre, fibres are oriented in all directions to innervate the various body tissue.

There is a pair of flame-cells or protonephridia, which are excretory organs and their ducts open to the outside between IV and V rows of epidermal plate cells.

During the differentiation of the miracidium within the egg shell, certain germinal cells become trapped in the parenchyma. These develop into germ balls by increasing in number and size and becoming enveloped, eventually giving rise to the next larval generation. Such germ balls are usually located in the posterior region of the body.

Miracidium swims about in water or moisture film, where cattle usually pass out their faeces. It dies in 30 hours unless it is able to reach intermediate host, which is always an amphibious snail of the genus *Limnea*. Intermediate host in case of *Fasciola indica* is *Limnea acuminatisata* while in the case of *Fasciola hepatica* intermediate host in *Limnea truncatula*. When larva enters the tissue of the intermediate host with the help of apical papilla, it reaches soft parts, such as mantle, pulmonary sac and hepato-pancreas. Penetration into the intermediate host is helped by the secretion of the penetration glands or other glands. On reaching the intermediate host, ciliated epidermal cells are cast off, while eyes and brain are lost. Larva swells up and changes in shape and size, so as to become a sac-like structure, called *sporocyst*. Germ-balls of miracidium do not disintegrate.

Sporocyst

The sporocyst is an elongated, hollow,sac-like larva. Its body wall consists of non-cellular cuticle followed by musculature and sub-epithelium. The brood chamber is hollow space in the centre of the sporocyst within which are found the germ balls that eventually give rise to the next generation. Sporocyst contains no alimentary canal, nervous system or reproductive systems. The protonephridia are present opening laterally by pores.

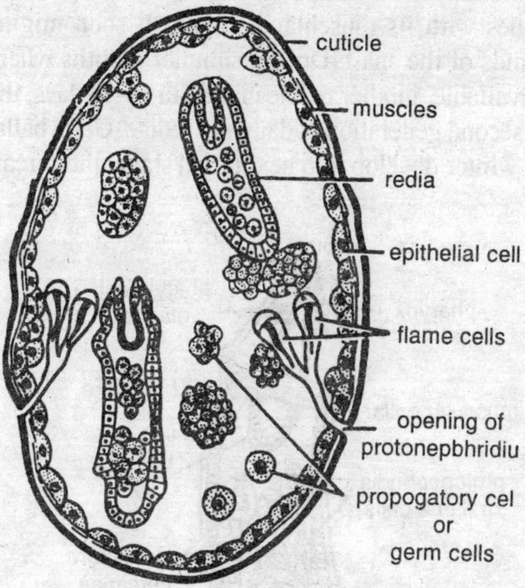

Fig. 3.17. *Fasciola*. Sporocyst.

The sporocyst moves about by muscular contractions in the host tissues, absorbing nourishment from them and causing them a considerable damage. Its germ balls develop into the next larval form, the rediae. Each sporocyst produces 5-8 rediae. Some of the germinal cells are again set aside within the rediae for the formation of the next larval forms.

Redia larva

The rediae emerge from the sporocyst by the rupture of body wall of sporocyst. Each radia is an elongated and cylindrical form, measuring 1.3-1.6 mm in length. It has a circular ridge, called *collar* near the anterior end and a pair of projections termed *lappets* near the posterior end. The body-wall consists of the usual cuticle, musculature and subepithelium. Unlike the sporocyst, each redia possess a mouth located at the anterior terminal of the elongated body. The mouth leads into a muscular pharynx,which in term leads a blind-sac, the gut. Numerous unicellular pharyngeal glands open into the pharynx.

Protonephridia divide further and form a much branched system. All the flame cells of one side open out through a common excretory duct. A birth pore is present, lateral to and in proximity of the mouth. Within the brood chamber are found germ balls.

Redia is a very active little creature. It moves about by the contractions of the body and lappets. It nourishes itself by sucking fluid and cells of the host tissues with its muscular pharynx. It soon migrates into the digestive glands of the snail. During summer months when the food is sufficiently available, mother rediae mature in 12-18 days, the germ balls give rise to a second generation of daughter rediae. Germ balls of daughter rediae in the winter develop into next larval stage, the cercariae.

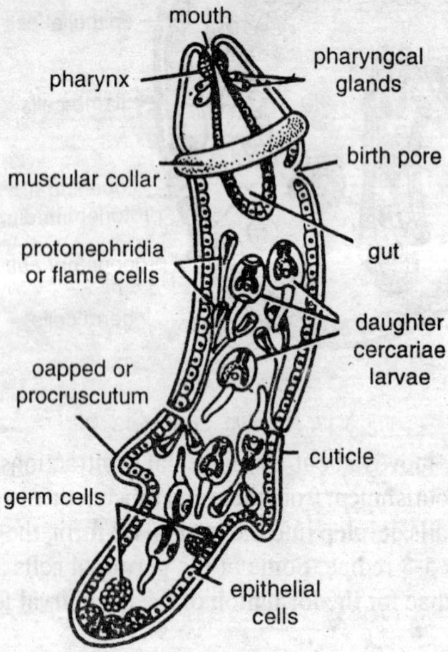

Fig. 3.18. *Fasciola.* Redia.

Cercaria larva

Cercariae leave the mother or rediae through their birth pores. From each redia 14-20 cercariae are produced. Cercaria is about 0.25-0.35 mm long and is an oval larva with a tail. The body wall consists of cuticle, musculature and subepithelium. On the anterior side, there is a mouth with a sucker, a suctorial pharynx, long oesophagus and a bifid intestine. A ventral sucker or acetabulum is present on the ventral side. A large number of dark, brown cells, called *cystogenous glands* are found beneath the cuticle. There is a pair of excretory ducts, each with several flame cells. These open near the hind end of the larva into a bladder that leads out through a pair of nephridiopores. A nervous system similar to that of adult has also appeared. Rudiments of reproductive organs develop from the germ cells near the hind-end.

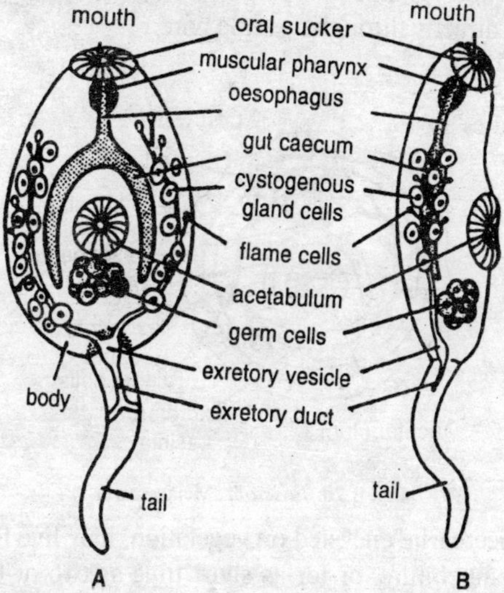

Fig. 3.19. *Fasciola.* A—Cercaria in ventral view, B—Cercaria in lateral view.

From the digestive glands of the snail, the cercariae pass into the pulmonary sac and then escape into the surrounding water. Cercaria swims about 5 minutes to an hour in summer. Afterwards, it settles down on some green leaves of water plants. Tail is cast off and the body becomes rounded and forms around itself a thick brownish cyst wall from the cystogenous glands. The encysted cercaria is now called metacercaria.

FACTORS AFFECTING DEVELOPMENT IN SNAIL

The rate of development in the snail host is a function of the environmental temperature. *Kendell* reported a curious effect of temperature on the development of *Fasciola*. In snails maintained at laboratory temperature, daughter rediae were not produced, the first generation of rediae giving rise to cercariae. On the other hand snails exposed daily to a period of low temperature i.e. 4-5 C°, daughter rediae were produced from rediae.

The state of snail nutrition also affects development. The effect of food availability on development in the snail host is probably the explanation for the observation that fewer larvae develop in small snails than in large ones.

Metacercaria

It is a juvenile fluke having a rounded form with a diameter of about 0.2 mm. It resembles cercaria larva but differs from later in the absence of tail, cystogenous glands and presence of thick cyst wall. Excretory bladder opens directly through a single pore.

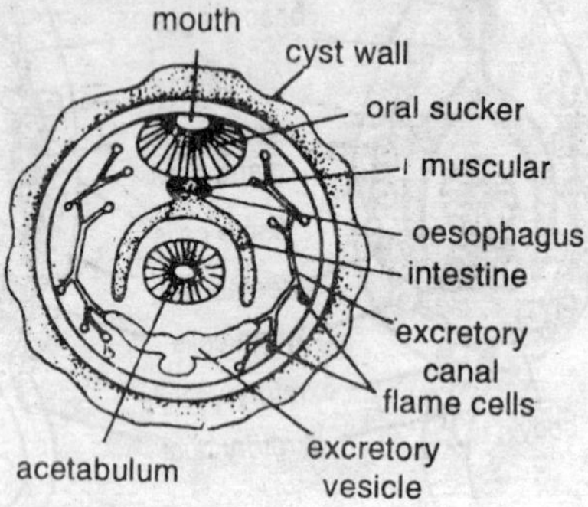

Fig. 3.20. *Fasciola.* Metacercaria.

The metacercaria, encysted on vegetation, may live for as long as a year at low temperature, or for as short time as two or three weeks at 25°C.

Infection of final host

The final excystment in the host requires an elevated temperature i.e. 38°C, low redox potential, a high carbon dioxide tension, and the presence of bile. The factors necessary for the excystment of *Fasciola* matacercariae seem to serve as stimuli to activate the worms, but the mechanism is not clearly understood. When metacercaria reaches the intestine of host , the cyst is digested and young fluke is liberated. The young worm penetrate the wall of gut and passes into the peritoneal cavity. In the body cavity the young worms seem to move about rather aimlessly until they encounter the capsule of liver. On contact with the liver, the flukes burrow into it, devouring liver parenchyma as they move. They grow in size within 5-6 weeks and finally settle in the bile duct. Each attains sexual maturity and begins to lay eggs 11-13 weeks after ingestion by the mammalian host.

Nature of life-history

The life-history of *Fasciola hepatica* is much complicated due to the occurrence of several larval forms in a series. Formarly, it was held that the rediae and cercariae develop by parthenogenesis from the propagatory or germ cells, which were believed to be the eggs (*Grobben*, 1882). This kind of asexual parthenogenetic reproduction by larval forms is known as *heterogamy*. This view was widely accepted for sometime and under its way, three terms were coined. These are *parthenitae* for sporocyst and rediae larvae, *adolescariae* for cercaria larvae, and *maritae* for adult flukes. This view, however, is now considered erroneous and has been given up.

It has been found that the germ cells present in the larval forms are not eggs. Instead they are diploid cells derived by the mitotic divisions of propagatory cell which is separated from the zygote at its first cleavage. Thus it can be said that they are parts of zygote. It, thus follows that one zygote produces several larvae. Thus, this process of reproduction in sporocysts and rediae has been looked upon as *polyembryony* by *Ishii* (1934), *Chen* (1937), *Rees* (1940) and *Cart* (1944). Since all the larvae resulting from a zygote are not formed simultaneously but at intervals, the phenomenon should be called delayed *polyembryony.*

Steenstrup (1942) and some other workers interpret polyembryony as a sexual multiplication and hold that there is a alternation sexual and asexual generation in life-history of *Fasciola.* But *Hyman* is of the view that it is a continuous ontogeny involving asexual multilplication in the larval stages.

Pathogenesis

Pathogenesis as a direct result of the flukes activities may be cither chronic or acute. *Acute fascioliasis* or *liver rot* occurs during the preadult migration of the flukes in the parenchyma of the liver, and sometimes other organs, for about 8 weeks. The most striking changes in acute fasciolasis are seen in the liver, where there are haemorrhagic flukes. The liver becomes swollen and in some instances, may become so swollen that the capsule ruptures. The by-products produced by worms are very toxic and cause *anaemia, diarrhoea, eosinophilia* etc.

Chronic fascioliasis occurs beyond 12 weeks, when the flukes have reached the bile ducts and are maturing sexually. Although milk and wool production looses are the usual results of chronic infections, deaths also occur.

The liver performs a vast array of essential functions in vertebrates, and it is axiomatic that liver damage will have a variety of detrimental effects. The adult fluke causes anaemia, haemorrhage and suppression in production of new red blood corpuscles. The result of liver damage is failure to gain weight in young animals, weight losses in older animals, reduced milk production and lowered fleece weights in sheep. The osmotic pressure of the blood is lowered and fluid is retained in the tissue spaces.

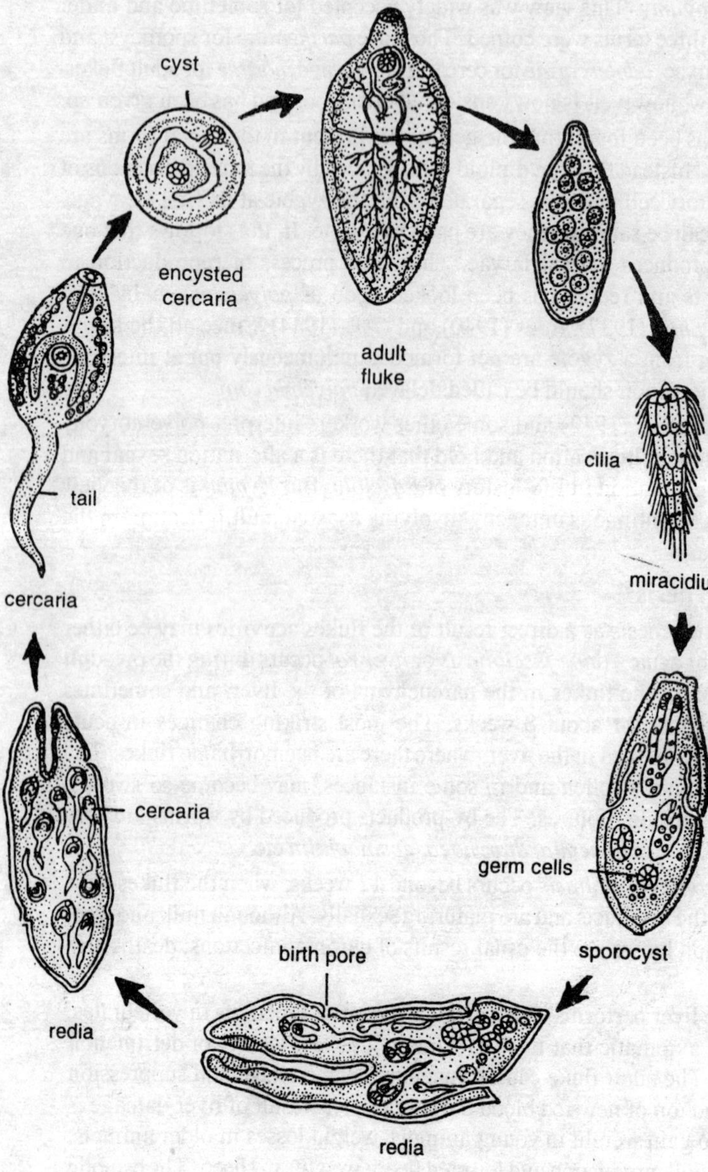

Fig. 3.21. *Fasciola*. Life-cycle.

Treatment

Carbon tetrachloride has been used in sheep at 500 mg/kg. Hexachlorethane was developed specially for use in cattles. Other drugs used are *emetine hydrochloride, filcin, tetrachloroethane*, etc.

Control

Control of *liver flukes* is effected in three major ways:

1. Snail control.
2. Treatment of infected sheeps.
3. Herd management through predicting out breaks of disease.

PARASITIC ADAPTATIONS IN *FASCIOLA*

The main changes that occur in the structure and life-cycle of a parasite, with the help of which it is capable of leading a parasitic life successfully, are called the *parasitic adaptations.* Since *Fasciola* is an endoparasite in the liver of sheep, it has undergone similar modifications under the stress of the parasitic mode of life which can be studied under two heads:

I. Structural modifications.
II. Modifications in the life-history.

I. Structural modifications

These modifications are of two types.

(A) Morphological
(B) Anatomical

(A) Morphological modifications

1. Body is dorsoventrally flattened leaf-like and is nearly triangular due to which the danger of injury to its body is minimized.
2. Adhesive organs or suckers are present, an *oral* and a *ventral* or *acetabulum.* These are muscular and help in attaching to the tissue of host.
3. External features simple and is not divisible into head, trunk and tail. Appendages are also absent.

(B) Anatomical modifications

1. Body is devoid of cellular epidermis, thick cuticle is present to prevent the animal from the effect of chemicals from host body. Cilia are absent in adult. Cuticle is provided with spinules for attachment.
2. The thickening of cuticle in the oral and ventral regions leads to the formation of two suckers.

3. The locomotary organs are lacking except in free living larval stages as adult *Fasciola* is an endoparasitic and it need not make excursions for food and safety.
4. The digestive system is incomplete. The *mouth* is anterior and terminal and surrounded by oral sucker. The pharynx is also muscular and suctorial. Both are adapted to suck the nutrient.
5. Intestine is highly branched so that digested food can be distributed to different parts of body. Anus is absent.
6. Circulatory and respiratory systems are absent. Anaerobic respiration takes place.
7. Excretory system is well developed. It consists of extremely branched excretory canals which terminate into flame cells.
8. Nervous system is poorly developed. Sense organs are absent.
9. Reproductive organs are extensively developed. The highly branched ovary and testes are capable of producing large number of gametes. Bisexuality offers full chances of fertilization.

II. Modifications in life-history

1. The production of eggs in a very large number, about 10 million, due to which the threat of survival of parasite is removed.
2. Eggs are shelled and each is provided with an operculum.
3. A secondary host is present for the dispersal of species.
4. Multiplication takes place by *paedogenesis* during larval stage.
5. Miracidia are free swimming larval stages because they have to search new hosts. The cercariae are encysted when out of the body of secondary host which ensures protection from external effects.

Revision Questions

1. Give a detailed account of structure of *Fasciola*.
2. Give an account of reproductive organs of *Fasciola*.
3. Describe the life history of *Fasciola hepatica*. Discuss its economic importance.
4. Describe parasitic adaptations in *Fasciola*.

5. What is a digenetic life-cycle ? Explain it with reference to the life-history of *Fasciola.*
6. Explain the life history of *Fasciola hepatica* with the help of diagrams only.
7. Draw labelled diagrams only :

   (a) T.S. of *Fasciola* through ovary, (b) Reproductive system of *Fasciola,* (c) Redia larva, (d) V.S. body wall of *Fasciola,* (e) Cercaria larva.
8. Write short notes on : (a) Flame cell, (b) Miracidium, (c) Redia, (d) Cercaria, (e) Polyembryony, (f) Ootype.

4

Schistosoma

The schistosomes are members of order Digenea, but are different from most members of order in that the sexes are separate. The term *Schistosoma* means split body and refers to the fact that the male has a ventral groove called *gynecophoral canal* in which female lodges. Adults of the family Schistosomatidae are found in the blood vessels of vertebrates, both birds and mammals. Three species namely *Schistosoma mansoni, S.japonicum* and *S. haematobium* are found in the blood of man and cause Bilharziasis or Schistosomiasis. No other disease caused by animal parasite causes as much harm to man as *Schistosoma* does with the sole exception of *Plasmodium* causing malaria. Schistosomiasis was discovered in 1779 in France. In 1851, *Bilharz* reported *S. haematobium*. *Harley* and *Cobbald* (1864) reported that some molluses are intermediate host. The molluscan hosts are species of *Tropicorbis*, *Biomphalaria* and *Australorbis*.

SYSTEMATIC POSITION

Phylum	-	Platyhelminthes
Class	-	Trematoda
Order	-	Digenea
Family	-	Schistosomatidae
Genus	-	*Schistosoma*

Distribution

The human schistosomes are distributed through some of the countries of the world at the lower end of the economic scale, or at least the parasites are found in relatively poor rural people. *S. haematobium* is found in Africa especially in Egypt and Sudan where up to 90 % of the rural masses may carry the infection. The intermediate host is the freshwater snail *Bulinus truncatus* and related species of *Bulinus*. *S. japonicum* is found in the Far East (China, Japan, Philippines, Thailand, Vietnam etc.). The mollusc host for *S. japonicum* is the species of the snail genus *Oncomelania*.

S. mansoni is found in many parts of Africa (including Egypt) and Tropical America including West Indies.

Habits and Habitat

They live in the blood vessels of the bladder, abdominal cavity etc. *S. mansoni* found in the portal blood of the host, usually in the smaller branches of the inferior mesenteric vein in the region of the lower colon. *S. japonicum* adults inhibit the branches of super mesenteric vein in the proximity of the small intestine. *S. haematobium* adults reach the vesicle, prostatic and uterine plexuses by way of inferior haemorrhoidal veins.

EXTERNAL MORPHOLOGY

The adult worms are beautifully adapted to life in blood vessels. The mature worms, male and female, live in copula (in almost permanent copulation).

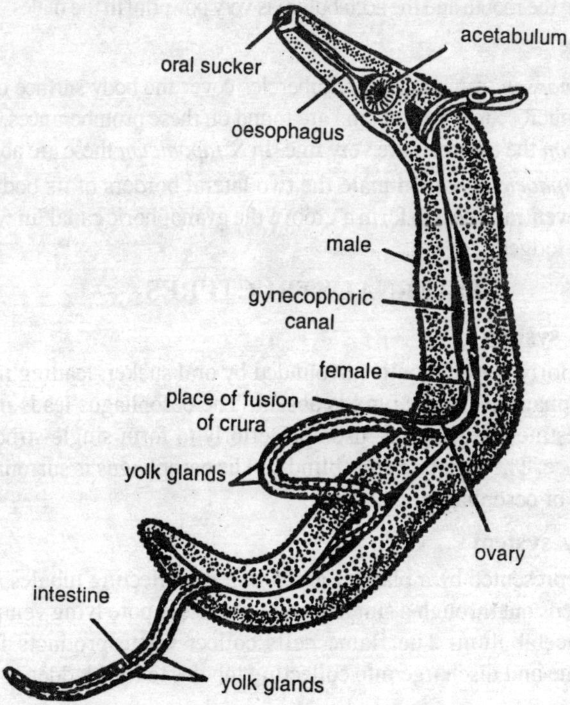

Fig. 4.1. *Schistoma*. Male and female.

Colouration

The colour of the body is greyish or pinkish. The alimentary canal presents reddish colour due to degenerating blood cells of host.

Shape and size

The body is thin, cylindrically elongated. The male are comparatively thicker and small than females. In *S. mansoni* the male, measuring 6.4-9.9 mm in length, is broader then the female. The female measures 7.2-14 mm in length. In *S. japonicum* the males may reach upto 22 mm in length while females 26 mm in length. In *S. haematobium* the adult males may become 15 mm long and the females 20 mm in length.

Suckers

Both male and female possessing two suckers each. Both the suckers are present and are located in the anterior portion of the body. The oral sucker surrounding the mouth and the acetabulum is very powerful in the males.

Tubercles

In *S. mansoni* minute sensory tubercles cover the body surface of the male, and microscopic tufts of cilia are found on these protuberances. In *S. haematobium* the tubercles are very fine. In *S. japonicum* these are absent.

Gyanophoric canal. In male the two lateral borders of its body are curved (on ventral sides) to form a groove the gyanophoric canal, in which the female lodges.

INTERNAL STRUCTURES

Digestive system

Anteriorly placed mouth surrounded by oral sucker, leading into a small oesophagus. The pharynx is absent. The oesophagus leads into a forked intestine which again fused posteriorly to form single structure for a distance. The intestine ends blindly. The oesophagus is surrounded by clusters of oesophageal glands.

Excretory system

It is represented by a bladder and a pair of collecting tubules. The bladder opens out through a single minute excretory pore lying ventrally near the acetabulum. The flame cells collect waste products from mesenchyme and discharge into collecting tubules to the bladder.

Nervous system

It is similar to other trematodes consisting of a circum-oral ring with cerebral ganglia and two lateral longitudinal nerve cords.

REPRODUCTIVE SYSTEM

Sexes are separate. Males are broader and enclose the females in the gynecophoric canals.

1. Male reproductive system

The male reproductive system consists of 4-7 testes which are spherical and are situated anteriorly. Each testis gives rise a minute duct known as *vas efferens,* All the *vasa efferentia* unite to form a common spermduct or *vas deferens*, which after a short distance dilates into a structure, the *seminal vesicle* just before opens into a *genital aperture* situated just below the acetabulum or ventral sucker.

2. Female reproductive system

Female reproductive system consists of a single *ovary*. From the posterior portion of ovary arises a coiled tube known as *oviduct*. The oviduct bends upwards and runs after a short distance to open into a *ootype*. The ootype in turn gives off a wide coiled tube, the *uterus*. The uterus is thin and long tube running between the two branches of intestine and opens into a *female genital pore* near the acetabulum. The ootype is surrounded by *Mehlis' gland*. *Vitelline glands* are situated in the posterior region of the body. From these glands a vitelline duct arises that runs along the oviduct and joins the latter before it opens into ootype. In *S. haematobium* seminal recepticle is absent.

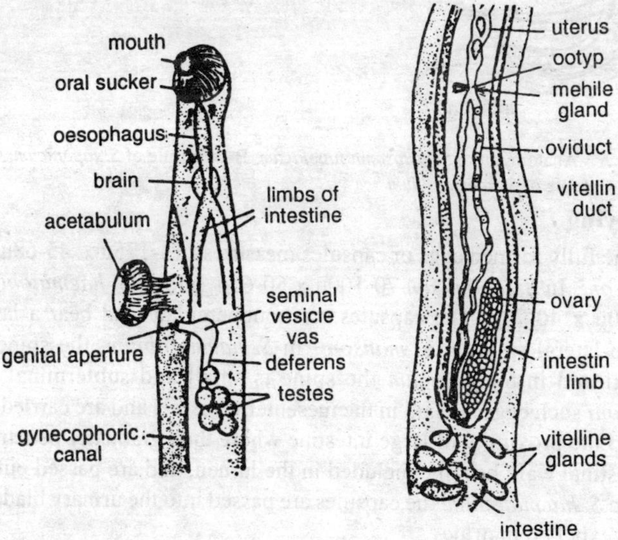

Fig. 4.2. A—Anatomy of male *S. mansoni*. B—Anatomy of female *S. haematobium*.

LIFE-HISTORY

Copulation and fertilization

The male and female adults in the definitive host move against the blood flow into small venules until they can move no further. *Fertilization* occurs when spermatozoa from the male are introduced into the female. Fertilization takes place in the *oviduct*. The fertilized eggs are encapsulated by the secretion of vitelline glands which also form yolk cells . In *S. haematobium* the uterus contains only 10-100 capsules while uterus of female *S. mansoni* contains only 14 capsules at a time.

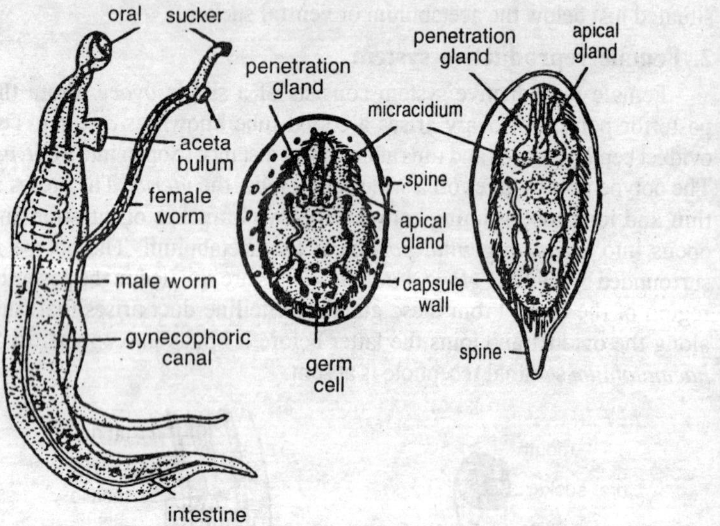

Fig. 4.3. A—Anatomy of *S.Schistosoma japonicum*, B—Capsule of *S. japonicum*, C—Capsule of *S. haematobium*.

Egg laying

The fully formed eggs or capsules measures 114-175μ x 45-68μ in *S. mansoni*. In *S. japonicum* 70-100μ x 50-65μ and in *S. haematobium* 112-170μ x 40-70μ. The capsules are nonoperculate and bear a large posterio-lateral spine in *S. mansoni*. In *S. haematobium*, the spine is terminal and in *S.japonicum* the spine is small and subterminal. In *S.mansoni* such eggs are laid in the mesenteric vessels and are carried by blood to the proximity of large intestine where they gradually penetrate the intestinal wall, become included in the lumen, and are passed out in feces. In *S. haematobium* the capsules are passed into the urinary bladder and are expelled in urine.

Female lays a single egg at a time. *Pesigan et al.* (1958) estimated that a female *S. japonicum* lays about 1200 eggs per day.

Miracidium larva

Eggs do not hatch in the urinary bladder or intestine of the host. The eggs of *S. japonicum* takes 9-10 days to develop mature miracidia while passing through the tissue and will live for 10 to 12 days longer if not expelled out, but may die in the tissue. *Miracidium* measures about 130μ in length and 60μ in breadth. The larva contains ciliated epithelium. Two pairs of penetration glands are present, one pair anterior in position other posterior. Miracidium also possesses a primitive gut, nerve centre and two pairs of flame cells. Germ balls are usually located in the posterior portion of the body.

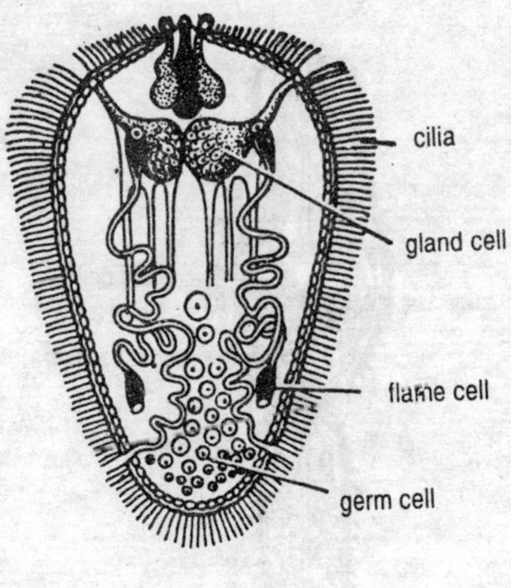

Fig. 4.4. *Schistosoma*. Miracidium.

When the eggs come in contact with water, at 28 °C the enclosed miracidia break out the shells by cracking them. The free-swimming miracidia finds and penetrates a pulmonate snail and transforms into mother sporocyst which is located in the foot region of the host. Free-swimming miracidium survive for upto 24 hours. It dies if not find its intermediate host within this single day of free life.

Sporocyst

In *S. mansoni* the sporocyst grows 1 mm in about 14 days. The sporocysts start producing daughter sporocysts which burst free of the mother sporocysts. These are differentiated from germ balls. The daughter sporocysts reach a length of 1.5 mm and produce forktailed cercaria from the germ balls. A single snail infected by a single miracidium may discharge about 3500 cercariae per day for sometime. The ceracariae are discharged during the forenoon on sunny days.

Cercaria larva

The ceracariae of *S. haematobium* and *S. mansoni* have a body of about 200μ long with a tail stem slightly longer and forks about 75μ long. For about two or three days the cercariae alternately swim and rest in water. *Cercaria* possesses an oral sucker and a ventral acetabulum. Mouth leads into the oesophagus which leads into a trifid intestine. Pharynx is absent. Five pairs of penetration glands are present-2 pairs of anterior *oxyphilic glands* and 3 pairs of posterior *basophilic glands*. If the cercariae fail to reach final host within two days they die.

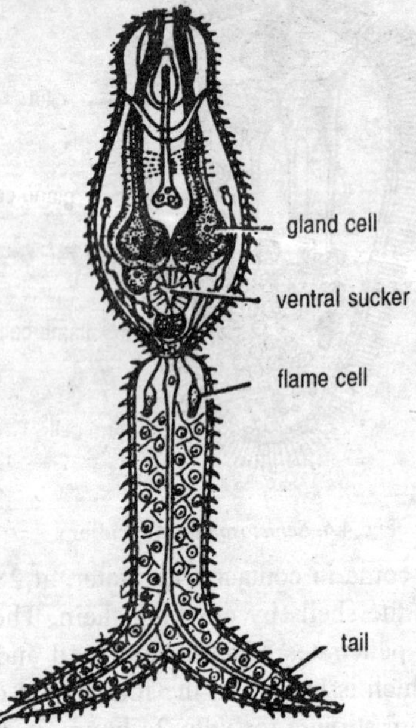

Fig. 4.5. *Schistosoma*. Cerceria.

Infection to the final host

If successful the cercariae burrow through the skin, using the histolytic and hyaluronidase-bearing secretion of their penetration glands just as the miracidia do. Skin penetration requires several minutes and may or may not be accompanied by a prickling sensation and subsequent dermatitis. If ingested with water the cercariae attach themselves to the mucous membrane of the mouth and similarly bore it. They leave their tail behind them. They find their way into the blood system and are carried via the heart to the lungs. Young *S.mansoni* accumulate in the lungs on the 2nd and 3rd days. By the 6th day they appear in number in the liver, where they are well established by the 15th day. Apperently cercariae feed only on portal blood, but once in the liver they grow rapidly. Migration to the mesentric veins begins about the 23rd day. The mating and egg production takes about 40th days. All larvae produced from one egg become adults of same sex. No female worms mature in a host, if the male worms are absent.

Pathogenicity

The injuries inflicted by schistosomes depends on the degree of infection and the prolonged period of infection. During migration of the eggs through tissue to reach the colon or urinary bladder tissue reactions in the form of inflammation, necrosis, connective tissue encapsulation, eventually scar formation take place. During migration of the eggs through the wall of the urinary bladder or colon, haemorrhage generally occurs. This is accompanied by *diarrhoea.*

In endemic areas, where reinfection occur, repeated penetration of colonal and bladder wall results in extensive scar formation, which prevents migration of eggs through these structures. As a results, the eggs, may transported to liver and rarely to brain or spinal cord. Whenever these aberrant eggs become lodged, local inflammation, fibrosis and necrosis occur. Cirrhosis of the liver is not uncommon. Swelling of liver and spleen is of common occurence. In the case of heavy infections, mature worms may migrate to the brain, lungs, uterus, oviduct and gonads. Their presence at these sites provoke pathological changes.

Acute cases of *schistosomiasis*, especially by *S.japonicum*, which is highly pathogenic, often leads to death.

Treatment

Treatment of schistosomasis has been difficult by the fact that relatively few drugs are available, their toxic effects are sometimes severe and they are active only against mature worms. Tartar antimonial gluconate and Stibocaptate have been used for years. Thioxanthones, Miracils, Lucanthone, Hycanthone, Niridazole imidazolidin z-one, etc. are also used.

Epidemiology and Control

The basic pattern of life cycles in the blood flukes was worked out in the first decade of the 20th century. With that information came the keys

Table showing differential characteristics of *Schistosoma* species of humans

	S.haematobium	*S.japonicum*	*S.mansoni*
Adults			
Male			
Length(mm)	10-14	12-20	6-12
No. of testes	4	7	6-9
Tegument	Scattered tubercles and blunt short spines	Minute spines	Numerous tubercules and large sharp spines.
Posterior union of intestinal caeca	Midbody	Posterior to midbody.	Anterior to midbody
Female			
Length (mm)	16-20	16-28	7-17
Ovary	Posterior to midbody	Midbody	Anterior to midbody
No. of eggs in uterus	20-100	50 or more	usually 1
Eggs produced by single female per day.	?	1400-3500	100-500
Location in host			
Primary host	Bladder and pelvic plexuses	Both superior and inferior mesenteric veins	Veins of small and large intestine near ilecocecal junction
Secondary host	Portal and mesenteric veins	Portal vein	Haemorrhoidal and hepatic veins
Prepatent period in weeks.	10-12	5-7	7-8
Eggs			
Length (μ)	112-170	70-100	114-175
Mean		82	
Width(μ)	40-70	50-65	45-68
Colour	Golden brown	Golden brown	Golden brown
Spines	Terminal	Subterminal knob	subterminal spine

to control. It was obvious that sanitary disposal of human excreta and reducing population of snails were the steps to take. As knowledge increased, additional techniques were developed, so that general methods used in attempting to control schistosomiasis are :

1. Snail control
2. Sanitary disposal of human excreta
3. Treatment of infected persons
4. Treatment or elimination of reservoir host
5. Prevention of penetration by cercariae
6. Immunization of persons at risk of infection
7. Prevention of infections in snails

Each method is directed towards the same end; to break the life cycle or interrupt the chain of infection. It all the snails could be removed from endemic area, the probably of infection would approach zero. If all definitive hosts were treated and maintained free of infection for a year, the cycle would be broken.

5

Opisthorchis sinensis

Opisthorchis sinensis is commonly known as 'Chinese Liver-fluke'. It was formarly known as *Clonorchis sinensis* and was described in 1875 from Chinese Hospital patient whose liver, on post-mortem, was found to be enlarged and turgid and from whose distended bile ducts many small trematode worms were recovered. Since that its epidemiology, therapy, morphology, prevention and life-cycle have been studied in details.

SYSTEMATIC POSITION

Phylum	-	Platyhelminthes
Class	-	Trematoda
Order	-	Digenea
Family	-	Opisthorchidae
Genus	-	*Opisthorchis*
Species	-	*sinensis*

Distribution

Opisthorchis is common in Japan, Korea, China, Hong Kong and Indochina. This distribution coincides with the distribution of the snail first intermediate host, *Parafossarulus manchouricus* and closely related hydrorbid snails (*Malek*, 1980). Nearly 100% of people in certain villages of China (*Komiya*, 1966) and over 65% of population of Hong Kong are infected.

Hosts and Location

Human is definitive host for *Opisthorchis*. Most fish-eating mammals and fish-eating birds can be infected, but dogs and cats are probably the most important reservoir hosts.

MORPHOLOGY

Opisthorchis is a smaller and narrower than *Fasciola* and not commonly related to it. Adult flukes are 10-25 mm in length and 2-4 mm in width. The body remains covered by rough cuticle. There are two suckers an *oral* and a *ventral* or *acetabulum.* The acetabulum is slightly smaller than the oral sucker and is located near the end of the anterior third of the body. *Genital pore* opens ventrally just infront of the acetabulum. An *excretory pore* lies at the posterior end.

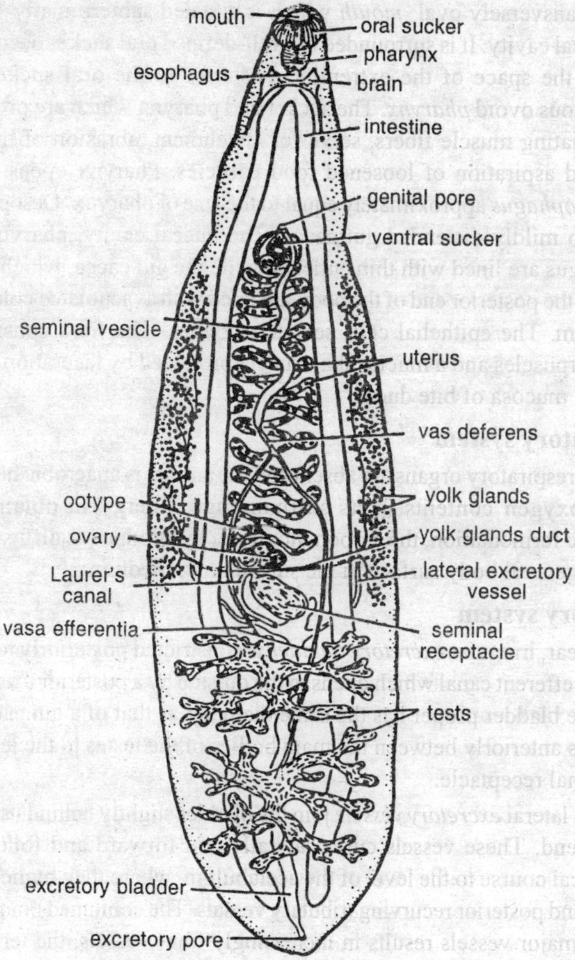

Fig. 5.1. *Opisthorchis sinensis.*

Integument and muscular system

A thin, minutely wrinkled cuticula which is devoid of spines overlies the hypodermis and the peripheral muscular layers of the body.

Parenchyma. The space between the organ system in the interior of the body is filled with a measenchyme called parenchyma. This tissue consists of irregular shaped cells between which is a matrix of fluid and interwining fibers.

Digestive System

A transversely oval *mouth* which is situated subterminally opens into buccal cavity. It is surrounded by well-defined oral sucker occupying most of the space of the extremity. Posterior to the oral sucker is a conspicuous ovoid *pharynx*. The sucker and pharynx which are provided with radiating muscle fibers, serve for attachment, abrasion of the bile ducts and aspiration of loosened food particles. Pharynx opens into a short *oesophagus* approximately equal to the size of pharynx. Oesophagus joins two mildly distended gut caeca. The buccal cavity, pharynx and oesophagus are lined with thin cuticula while the gut caeca, which reach nearly to the posterior end of the body,are lined with avacuolated columnar epithelium. The epithelial cells secrete enzymes. The food consists of blood corpuscles and a mucin-like exudate produced by laceration of the inflamed mucosa of bite duct.

Respiratory system

The respiratory organs are absent and the animal is anaerobic because of low oxygen contents in its surroundings. Energy in obtainedby anaerobic fermentation, the carbon dioxide, a by-product, is diffused out through general body surface in the surrounding medium.

Excretory system

A clear, irregular *excretory bladder* is constricted posteriorly to form a narrow efferent canal which opens to the outside by a posterior *excretory pore.* The bladder proper has the same diameter as that of a gut caecum. It follows anteriorly between the main bodies of the testes to the level of the seminal receptacle.

Two lateral *excretory vessels* join the bladder slightly behind its blind anterior end. These vessels curve laterally and forward and follow an extracaecal course to the level of the acetabulum, where they branch into anterior and posterior recurving tributary vessels. The continued branching of these major vessels results in increasingly finer vessels, the terminal units of which are the flame cells.

Nervous system

A bilobed brain with medial commissure and anterior posterior nerves is located in the dorsal oesophageal region.

Reproductive system

Opisthorchis is hermaphrodite trematode possesses well-defined reproductive organs and a common genital pore.

Male reproductive system

Testes are posterior,branched and lying one behind the other. From the main body of each testis arises a thin duct the *vas efferens*. This duct proceeds forward to a point slightly anterior to the ovary where it joins its mate from the other testis to form a vas deferens enlarges into a sperm-filled *seminal vesicle*. A narrow ejaculatory duct connects the vesicle to the genital atrium which opens at the ventral genital pore. A cirrus sac prostate gland and cirrus all are absent.

Female reproductive system

A single branched *ovary* is present in the median line slightly anterior to an oblong sperm-filled *seminal receptacie.* The seminal recepticle lies just ahead of the anterior testis. Anteriorly the seminal recepticle gives off a short duct that joins Laurer's canal and which terminally joins a short *oviduct. Laurer's canal* curves posteriorly around the seminal recepticle and also runs abruptly dorsally to open by a small aperture, the *opening of Laurer's canal*. This opening lies between the anterior testis and seminal recepticle near the middle. The oviduct after receiving the seminal recepticle, proceeds forward a short distance and receives a common *vitelline duct. Vitelline glands* are small and numerous located in the middle third of the body. Two lateral vitelline ducts arise from the middle of transverse vitelline duct. The oviduct receiving all these ducts (i.e. median vitelline duct, Laurer's canal and seminal recepticle) then penetrates an *ootype* which is surrounded by *Mehlis' glands.* The *Mehlis' glands* are loosely organized and composed of small acinus shaped cells. A very conspicuous *uterus* arises from the ootype anteriorly and proceeds as an irregularly coiled duct to the *genital atrium*. The uterus becomes larger in diameter anteriorly because of the massing of eggs.

Fertilization

Mutual cross-fertilization occurs after copulation between two flukes. The ova liberated from ovary are fertilized by the sperms from the seminal recepticle. In the ootype the fertilized egg and yolk are enclosed in a thin walled, oval, operculate shall which is often provided with a comma-shaped posterior appendage. The egg gradually acquires a dark brown colour and proceeds with cleavage until a fully formed *miracidium* is formed. The embryonated capsules, pass to out side through the genital pore and are carried by the bile to the host's intestine, they are finally eleminated with the feaces.

LIFE-HISTORY

Man is the primary host besides snail and fish are two intermediate hosts:

Miracidium and sporocyst

As stated above the eggs when laid contain fully mature miracidia. Each ovoid miracidium bears long, motile cilia on an outer ectodermal sheath. The apical region bearing no cilia. In the apical region there is no apical spine and an apical gland opening. On the dorsal side there is a single excretory pore. An apical histolytic gland and a large, ventral, sausage-shaped secretory gland with a narrow duct opening besides the spine. About 25 germ balls are present, that fills the miracidial body space. Two dorsally placed flame bulbs and their ducts comprise the excretory system. Nervous system is situated ventral to the apical gland.

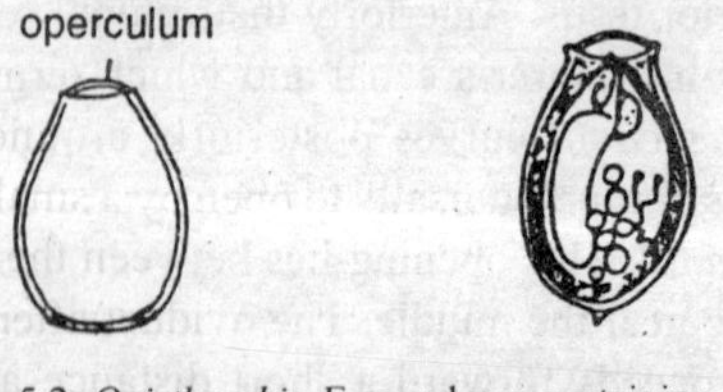

Fig. 5.2. *Opisthorchis.* Egg and egg containing miracidium.

Miracidium do not hatch free in nature, rather they are released from the eggs only after ingestion by appropriate snail host. The miracidium than actively penetrates the gut wall into the peri-intestinal blood sinus. Now it becomes a thin walled *sporocyst* produces the redial generation which migrates to the hipatic blood sinuses.

Redia

A mature redia is elongated in appearance. Its anterior end is rounded whereas posterior end is somewhat pointed. The larva is internally lined with sub-epithelium. A birth pore is absent. An anterior pharynx and a brownish sacculate gut extending about half-way posteriorly, are present. Several germ balls are present which develop synchronously into cercariae. About 6-8 cercariae are liberated by the rupture of redial wall.

Cercaria

The long-tailed, oculated cercaria emerges from the snail host into the water and swims vigrously in the search of another host i.e. a fish. The tail is provided with muscle-fibers and bears thin, lateral, transparent elytra. Two bean-shaped eye-spots are located near the anterior end of body. An oral-sucker, pharynx and acetabulum are present but oesophagus

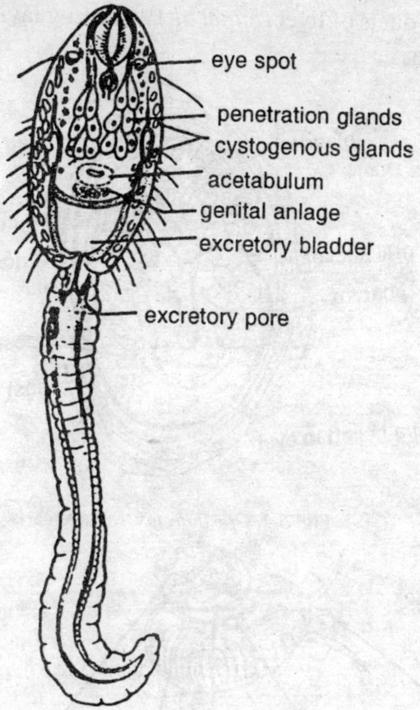

Fig. 5.3. *Opisthorchis*. Cercaria.

and gut caeca are not formed yet. Seven pairs of unicellular glands occur in the middle of the body. Their ducts open near the mouth. Dorsal and ventral peripheral mesenchymal cells are packed with cystogenous glands. These glands help in the formation of cyst of metacercaria. Excretory system comprises a triangular excretory bladder with anterior-lateral collecting and a bifurcated posterior caudal excretory tubule.

Metacercaria

On coming into contact with the fish, member of Cyprinidae family, the cercaria attaches to the surface, discards its tail, and creeps under a scale into the fish. A cyst wall is formed by the material secreted by cystogenous glands. The cyst is thin, tough, and spherical. Organ formation continues e.g., the gut caeca etc.

Infection to the primary host

Primary host i.e. man ingests uncooked fish containing metacercaria. Metacercarial cyst wall dissolves in the duodenum or intestine (*Gibson and Sun*, 1971) and juvenile flukes migrate up the common bile-duct into the small bile ducts of liver (*Sunet al*,1968). Egg laying begins 3-4 weeks after infection.

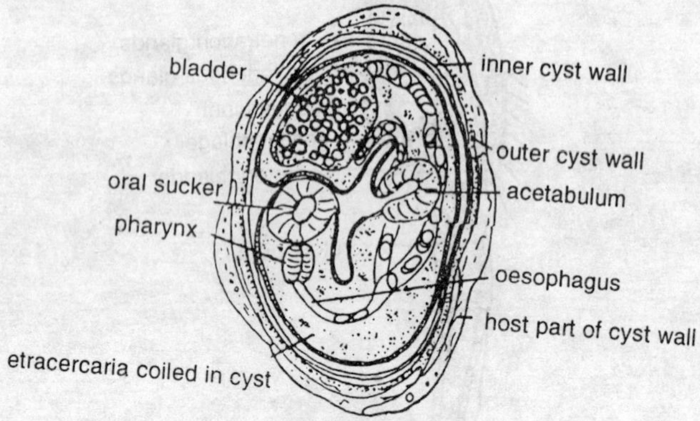

Fig. 5.4. *Opisthorchis.* Metacercaria.

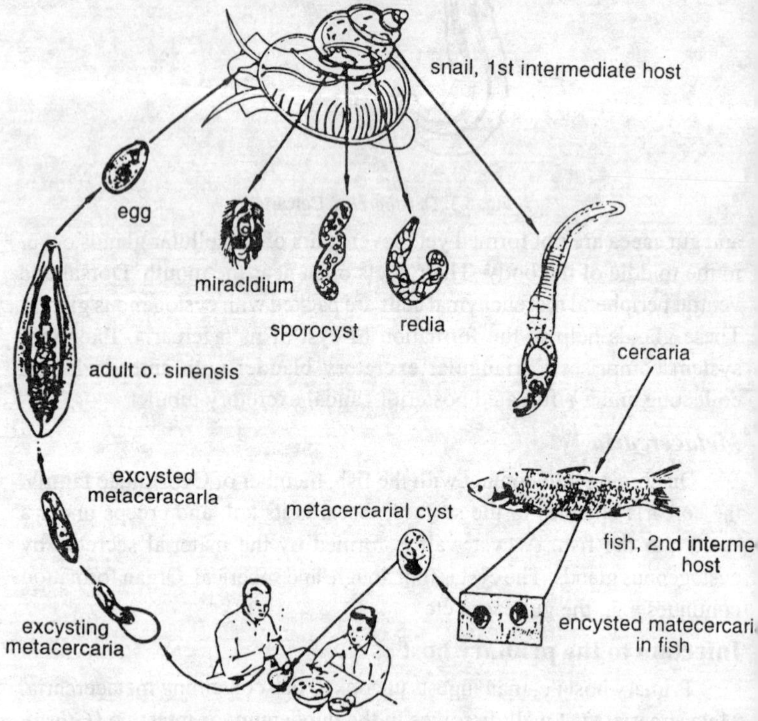

Fig. 5.5. *Opisthorchis. Life cycle.*

Effects on the host

Symptoms are classified as mid, modrate or severe, depending upon the burden of worm, length of infection and health of host. Light infections may be asymptomatic. Heavy infections are characterized by chills and fever, followed by liver enlargement and tenderness, sometimes with splenomegaly (Spleen enlargement). Chronic heavy infections cause cholecystitis (inflammation of gall bladder) and hepatitis (inflammation of liver). Modrate infection causes diarrhoea, edema and some hepatomegaly (enlargement of liver). The disease is termed *opisthorchiasis* or *clonorchiasis*.

Treatment

Hexachloroparaxylol, Hetol and Paraziquantal are used. The first two appear to be useful.

Control

All the fishes, used as food, should be properly cooked to avoid infection.

Molluscicides to control the snail (first intermediate host) are generally not used because they are harmful to important fresh water fishes (*Malek*, 1980). Animal and human feaces are important fertilizers for fish ponds in the Orient. Faeces must be properly composted. In addition, snail populations have significantly reduced by removal of water plants when snails lay their eggs, there by permitting fish more easily to feed on larval snails.

Revision Question

1. Describe the structure and life cycle of Chinese Liverfluke ?
2. Describe the habits and habitat, structure and control measures of *Opisthorchis sinesis*. ?
3. Draw a neat and well labelled diagram of *Opisthorchis* ?

6

Taenia solium

Taenia solium belongs to class Cestoda. The members of this class are commonly known as *Tapeworms*. All are endoparasites in the alimentary canal and associated ducts of various vertebrate hosts. During their life cycles one or two or more intermediate hosts, vertebrate and invertebrate, are required in which the tapeworms undergo a phase of development. The cestodes differ, however, from the members of other two classes (Turbellaira and Trematoda) in the complete absence of a digestive tract.

The present text mainly related to *Taenia solium*, the pork-tapeworm of man.

SYSTEMATIC POSITION

Phylum	-	Platyhelminthes
Class	-	Cestoda
Subclass	-	Eucestoda
Order	-	Cyclophyllidea
Genus	-	*Taenia*
Species	-	*solium*

Habits and habitat

Taenia solium or porkworm is commonly found in those parts of the world where *pork* is eaten as food without properly cooking. It is found as an endoparasite in the gut of vertebrate hosts. One host is man and another is pig. Formerly, it was very common intestinal parasite in Europeans, especially in some places of Europe. It is rare in many parts of the world viz. India, Philippines, North Amercia. It is not at all found in Jewish and Mohammedan countries as in those countries eating of the flesh of pig is considered as an evil deed.

EXTERNAL MORPHOLOGY

Shape, size and colour

The body is elongate and flattened dorsoventrally like a ribbon or tape. The body is narrow anteriorly and gradually broadens laterally towards posterior end.

The size of adult worm varies from 3-5 meters but a few are recorded to attain a length of about 8 meters.

The body is usually opaque white in colour but creamish, yellowish or greyish colouration is also common.

Division of body

The body of *Taenia* is distinguished into three regions.

1. Scolex
2. Neck
3. Strobila

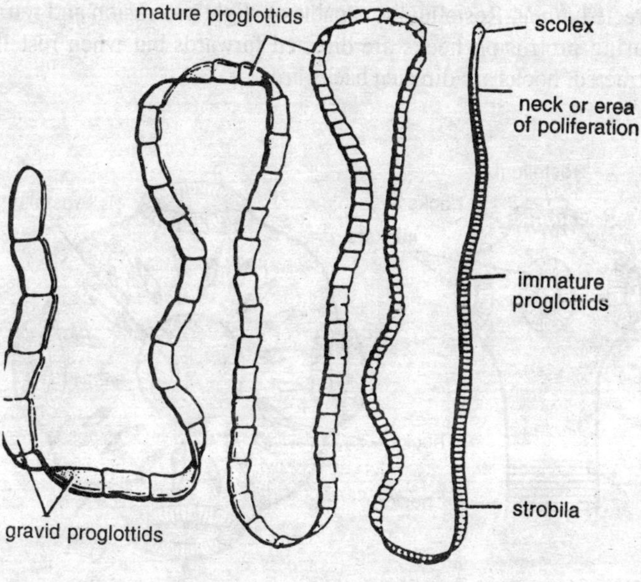

Fig. 6.1. *Taenia solium.* Entire.

1. Scolex

The scolex is the holdfast organ, which is generally considered the anterior end. There are two schools of thought concerning the interpretation of the structure of tapeworms. Some consider a single scolex and connected proglottids to be a single animal. Others consider each proglottid to be an individual - hence, a single tapeworm is a colony of many individuals. Some helminthologists argue that the scolical end of the cestode is actually the posterior end. However, *Hyman* (1951) and *Chang* (1964) are not convinced with this arguement that scolex is the posterior end. Furthermore, the more recent literature indicates the scolical end should be considered the anterior end.

Scolex is a pear-shaped structure, but is four-sided at the posterior part. It possesses a small terminal cone called the *rosetellum*. Around the base of the terminal cone or rostellum is a double row of curved, pointed, chitinous hooks, which are usually 28 in number but are sometimes between 22 to 32. Hooks are of two types, long and small. Both long and small hooks are arranged alternately. Long hooks are approximately of 180μ and short ones are of 130μ dimension.

Each hook consists of a *base* or *guard* by which it is fixed, a blunt projection or handle directed towards the apex and a conical outwardly directed *blade*. Rostellum is capable of slight protrusion and retraction. During protrusion hooks are directed forwards but when rostellum is retracted, hooks are directed backwards.

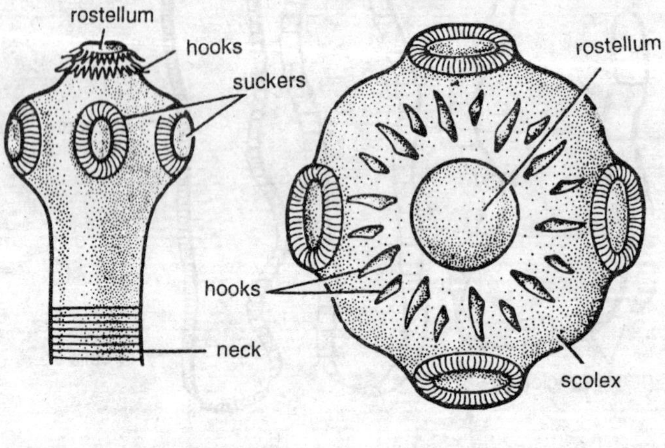

Fig. 6.2. *Taenia solium*. A—Scolex magnified, B—Frontal view.

The mouth is absent in scolex. The term "head" frequently used for the scolex is, thus, inappropriate, because the head is essentially that part of the body where organs for *perceiving* and *catching* food are aggregated. It measures about 0.6-1 mm in diameter.

2. Neck

It is situated immediately posterior to the scolex. This is an unsegmented, poorly differentiated area that is generally narrower than the scolex and the strobila proper, and is the continuously differentiating proglottids that is why called the *budding zone* or *area of proliferation* that gives rise to immature proglottid .

3. Strobila

The strobila, which constitutes the main bulk of body and is made up of a chain of proglottids. There are about 800-1000 such *proglottids.* Each proglottid bears a complete set of genitalia. This linear repetition of genital organs is termed as *proglottisation.* Since the proglottids are budded off from the neck region in an orderly succession, the youngest proglottids are towards the neck and the oldest proglottids are posterior most.

Depending upon the development of genital organs, the proglottids are distinguished into three types :

(i) *Immature proglottids.* These are young, undifferentiated proglottids next to the neck region. The reproductive organs are either absent or in stages of development. Their width is more than the length. They comprises about 200 anterior proglottids.

(ii) *Mature or reproductive proglottids.* The mature proglottids are sexually mature consisting of a male and a female set of reproductive organs. They are usually as long as broad and are found in the middle of the body. They comprise about 450 proglottids. Taenia is protandrous, about 100 or 150 anterior mature proglottids contain only male reproductive organs and remaining 300-350 contain both, male and female reproductive organs.

A mature proglottid is a complete reproductive unit in its own right and may undergo self or cross fertilization.

(iii) *Gravid or ripe proglottid.* These are found in the posterior part of the body. These are characterised by the possession of a large, highly branched uterus filled with developing embryos. The rest of the genital organs have already disappeared from them after performing their function. They are longer than broad. The posterior most proglottids are cut from the body and are expelled out of host body along with its faeces. This process of separation of proglottids is known as apolysis.

Body-wall

The body wall of cestodes is made up of several layers but lacks a cellular epidermis. It consists of (i) cuticle (ii) Basement membrane and (iii) Musculature.

(i) Cuticle

The outer most layer of the body surface is called the *cuticle* or *tegument.* This layer is more or less transparent. noncellular. and without visible pores. The cuticle consists of two layers. (a) outermost hair-like or fringe-like *comidial layer* and (b) a thick *homogenous* layer. The cuticle is secreted by modified *mesenchymal cells*. These cells lie beneath the musculature and have long processes extending up to the cuticle.

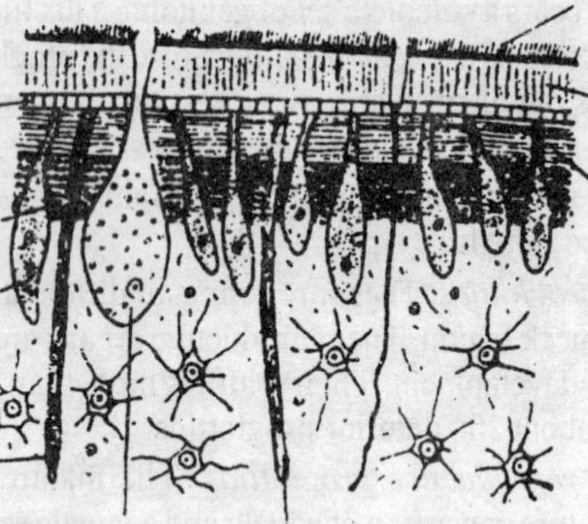

Fig. 6.3. *Taenia solium*. V.L.S. body wall seen under light microscope.

When studied under electron microscope, minute projections known as *microtriches* are present on the cuticular surface. *Rothman* (1963) suggested that the distal portion of each microthrix, which is solid, is concerned with two functions (a) It may serve as a means of resisting the intestinal current, since the body surface always in contact with the *microvilli* of the striated border of the cell lining of small intestine of host (b) it may serve to agitate the microhabitat in the vicinity as the worm moves, and thus stirring up the intestinal fluids so that nutritive materials as well as waste products are always in a state of flux. The proximal position of microthrix is medullated and could very well serve as sites of absorption. In addition to microtriches, ultramicroscopic pore canals that communicate with the parenchyma are present in cuticle. These may be the sites of absorption also. Mitochondria and endoplasmic canals are also reported in the cuticle. Thus, it is apparent that this outer body layer is a dynamic structure rather than a mere covering.

(ii) Basement membrane

Immediately beneath the cuticle is the basement membrane. This noncellular connective tissue layer contains vacuoles and granules. The protoplasmic layer beneath the basement layer is thick.

(iii) *Musculature*

Immediately beneath the basement membrane is a thin layer of *circular muscles* followed by thicker layer of *longitudinal muscles* and *vertical* or *dorsoventral muscles.*

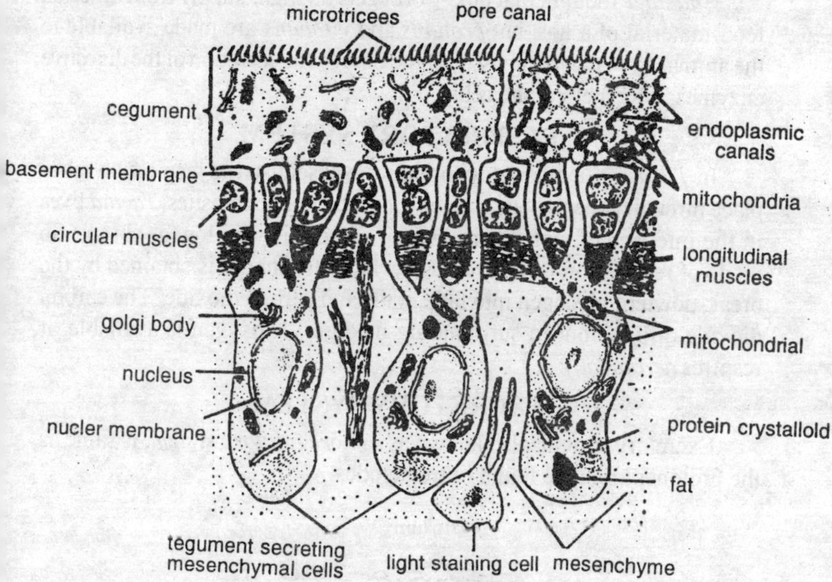

Fig. 6.4. *Taenia solium.* V.L.S. body wall seen under electron microscope.

Mesenchyme or parenchyma

The space enclosed by the body wall, except for the occupied by the reproductive organs, osmoregulatory structures, muscle fibres and nervous tissue, is filled with a spongy type of tissue, the *parenchyma.* Certain syncytially arranged cells form a network. In living animals, the space between the cells are filled with fluid. Numerous rounded calcarious bodies composed of concentric layers of calcium carbonate are present in mesenchyma secreted by *lime cells.*

Except at the two sides in each segment the parenchyma is divided into (i) outer *cortical* part and (ii) inner *meduallary* part by circular muscle fibers of the mesenchymal musculature.

The parenchymal cells, and to some extent the spaces, serve as sites for the storage of glycogen.

Nutrition

In *Taenia* there is no alimentary canal and as it lives in the intestine of host, there is totally absence of alimentary canal in all the stages of its life-cycle.

Its chief source of nourishment is obtained from the intestine of host. These are amino acids, glucose, glycerol etc.

Chandler thought that tape-worm gets its sugar supply from digested food-material of a host but *proteins* and *vitamins* are made available to the animal from the living tissue of the host by the secretion of the digestive enzymes, which effect the cells.

RESPIRATORY SYSTEM

In *Taenia*, the respiratory organs are absent so the respiration takes place through general body surface. Like other endoparasites, *Taenia* lives in the intestine of man where oxygen content is very low or absent. So mode of respiration is *anaerobic* in which the energy is obtained by the break-down of glycogen into fatty acids and carbon dioxide. The carbon dioxide diffuses out in surrounding medium. If oxygen is available, it respires *aerobically*

EXCRETORY SYSTEM

Excretory system, also known as *osmoregulatory system*, is same as the protonephric type found in tremetodes.

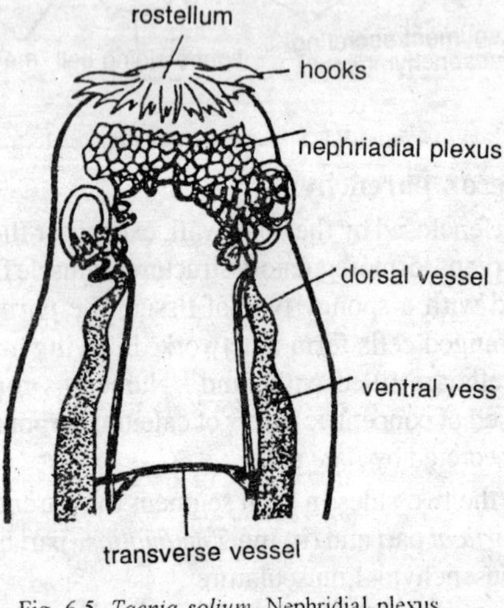

Fig. 6.5. *Taenia solium*. Nephridial plexus.

Four main collecting canals traverse the entire length of the strobila. The two *ventral canals* are ventrolaterally located, and the two *dorsal canals*.are dorsolateral. All these canals are situated in the periphyrangeal zone of the medullary parenchyma. A single *transverse canal* connects the two ventral canals at the posterior end of each proglottid. The ventral vessel carries water away from the scolex, the dorsal vessels toward it. Within the scolex, all the four longitudinal canals may be joined by a network of canals forming *nephridial plexus*.

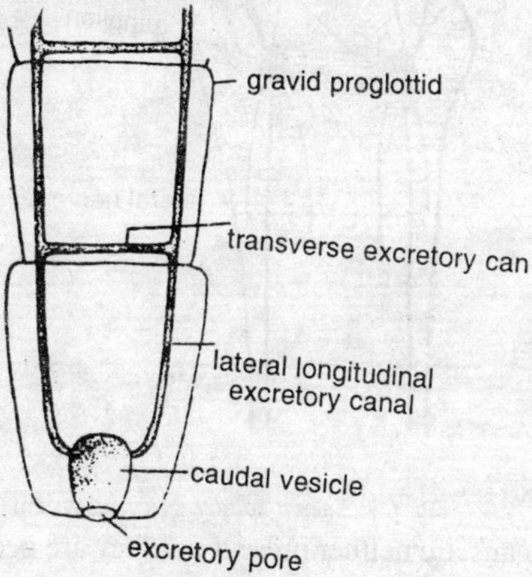

Fig. 6.6. *Taenia solium*. Caudal vesicle in last segment.

Along the length of ventral canals, a series of *secondary tubules* arise that may in give rise to tertiary tubules. At the free end of terminal tubules are *flame cells*. The flame cells are generally arranged in a group of four. A flame cell is of irregular shape, with granular cytoplasm and nucleus. The 'flame' is actually a group of cilia that arise from a concave basal plate located near the cell nucleus. The cilia are enveloped in the funnel-shaped enlargement of the free end of the tubule. Waste products collected through the flame cells is passed down the tubules into the *main canals*.

In the young worm, there is an *excretory vesicle* in the posterior most proglottid, into which the ventrolateral canals open. However, as the neck region produces more proglottids, the older proglottids are pushed further back, and eventually the proglottid containing excretory vesicle is detached by the process of apolysis. Thus, in most older animal the posterior end of the tubes open independently to the exterior.

NERVOUS SYSTEM

A thick band in the scolex with two ganglia constitute the *brain* or *central nervous system*. From the brain few nerves are given out anteriorly which innervate the scolex. Three pairs of longer nerves are given out towards the posterior region. Out of these only one pair, the *lateral nerve cords* extend through the strobila passing just outside the excretory canal in each proglottid.

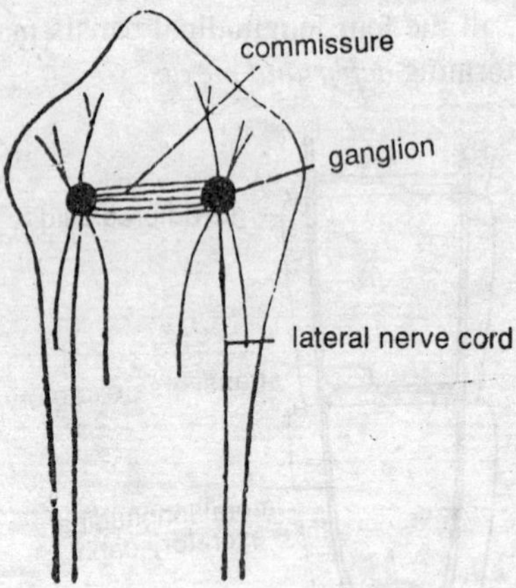

Fig. 6.7. *Taenia solium.* Nervous system.

Sense organs are neither present nor they are necessary. Free nerve endings are found in the body wall and detached proglottids for some time show independent muscular movement.

REPRODUCTIVE SYSTEM

Taenia is a *hermaphrodite*. Reproductive organs are usually found in the proglottids after 200th partition. Some anterior proglottids are functionally male. Middle proglottids are both male as well as female. Proglottids at the posterior end are filled with enlarged *uteri*.

Structure of the proglottids in the middle region is as follows:-

Male Reproductive System

The *Male Reproductive System* consist of the following organs:

1. ***Testes.*** They are numerous small, rounded bodies situated in the medullary part of the parenchyma. There number is about 15-200 and are distributed throughout close to the dorsal surface of the proglottid. Some workers are of the opinion that the numerous rounded follicles, in fact, constitute a single and highly sub-divided testis. However, the number of testes is of taxonomic value in tapeworms.
2. ***Efferent ducts.*** From each testis arises a fine ductule or capillary known as *vas efferens*. All the *vasa efferentia* unite to form a *vas deferens*.

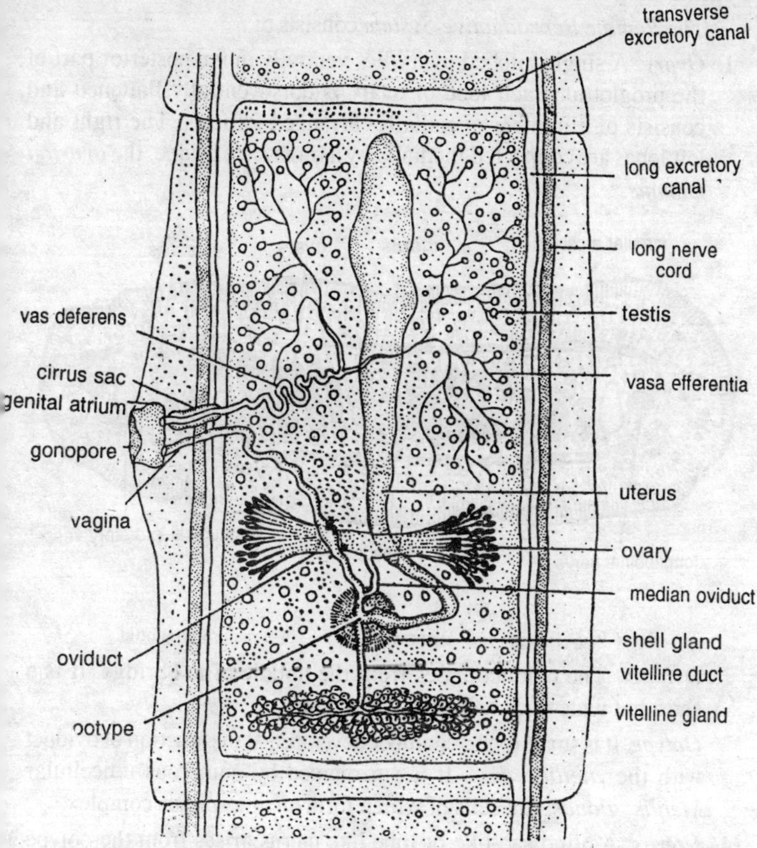

Fig. 6.8. *Taenia solium.* Mature proglottid showing reproductive system.

3. ***Vas deferens.*** It is a long tube, arising from the mid-point of the proglottid. It runs transversely to the left or right to open into cirrus. In some cestodes the vas deferens enlarges to from the seminal vesicle for the storage of sperms. But in *Taenia* no such enlargement is seen.
4. ***Cirrus or penis.*** The distal end of vas deferens in modified into a protrusible cirrus. The cirrus is armed with bristles, hooks or spines. It is eversible through a male genital pore into a cup like *genital atrium* of the *genital papilla*.
5. ***Cirrus sac.*** Cirrus is surrounded by a kind of muscular sac known as *cirrus sac*. Its function is to erect penis during copulation.

Female Reproductive System

The *Female Reproductive System* consists of :

1. ***Ovary.*** A single bilobed ovary lies ventrally in the posterior part of the proglottid. Each lobe of ovary is dorsoventrally flattened and consists of a number of radially- arranged follicles. The right and left lobes are connected medially by a transverse bridge, the *ovarian isthmus*.

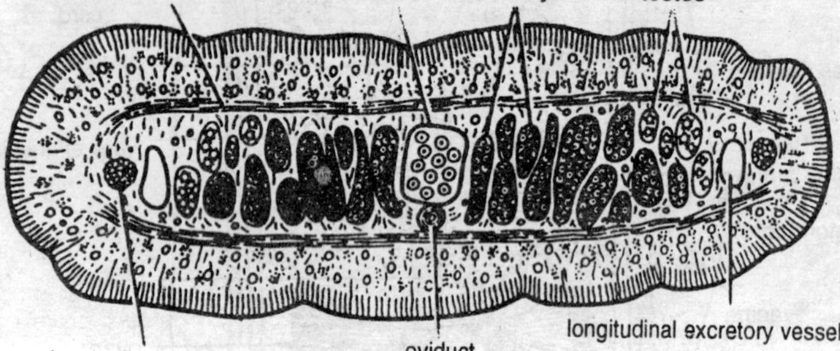

Fig. 6.9. *Taenia solium*. Diagrammatic T.S. of a mature proglottid.

2. ***Oviduct.*** The oviduct arises from the middle of the bridge. It is a short but wide duct and joins the ootype.
3. ***Ootype.*** It is small round chamber developed at the junction of oviduct with the *vitelline duct*. It is surrounded by numerous unicellular *Mehlis' glands*, the cells of which form an amorphous complex.
4. ***Uterus.*** A blind sac-like or tube like uterus arises from the ootype and runs forward in the segment. In gravid proglottids the uterus enlarges and gets branched to occupy the whole of proglottid. It remains filled with the fertilized ova or developing embryos.
5. ***Vagina.*** The narrow tubular vagina arises from the receptaculum seminalis and runs obliquely outwards to open into the genital chamber by female reproductive opening.

   The seminal receptacle or receptaculum seminalis is a small wide tube connected on one hand with the ootype by a narrow *spermatic duct*, and on the other with *vagina*.
6. ***Vitelline gland.*** The vitelline gland is a compact, elliptical mass of numerous follicles situated posterior to the ovary. A single short median duct, *median vitelline duct* arises from the gland and opens

into the ootype. The secretion of the vitelline duct is rich in yolk and forms a covering of yolk and shell around the fertilized egg.

7. ***Mehlis' glands.*** These are numerous minute unicellular glands situated around the ootype and emptying into it. The function of the secretion of Mehlis' glands is to ease the passage of shelled eggs into the uterus.

LIFE CYCLE

Fertilization

Fertilization occurs by introducing the cirrus into the vagina of the same proglottid and sperms are stored in the receptaculum seminalis, which later on fuse with eggs in uterus. Thus in *Taenia* self-fertilization takes place. Sometimes, the anterior mature proglottids, devoid of female genitalia, can enter the stage of copulation only with the posterior mature proglottids with fully developed female genital organs. Fertilization, following copulation between two different proglottids, is sometimes termed *cross-fertilization* to distinguish it from that occuring between gametes of the same proglottid (self-fertilization).

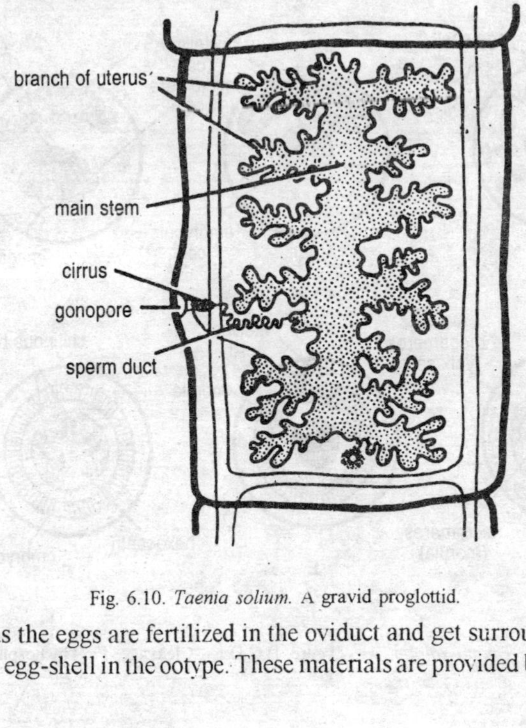

Fig. 6.10. *Taenia solium.* A gravid proglottid.

Thus the eggs are fertilized in the oviduct and get surrounded with yolk and egg-shell in the ootype. These materials are provided by vitelline gland. The capsulated eggs enters the uterus and are collected there. The uterus enlarges in size, gets branched and occupies the whole space. The eggs are small measuring 40μ (micron) in diameter.

Cleavage

The *zygote* or *egg cell* undergoes cleavage when the egg capsule is inside the uterus. The first unequal division results in a larger *megamere* and a smaller *embryonic cell*. The megamere divides further and froms several similar megameres, while the embryonic cell divides repeatedly producing two types of embryonic cells, larger *mesomeres* and smaller *micromeres*. Thus, three types of cells result from the zygote; small *micromeres*, medium *mesomeres* and large *megameres*.

Onchosphere

By repeated division of micromeres a solid ball of cells, the *morula* is formed. The megameres divide, redivide and become filled up with yolk from the yolk cells, and surround the embryonic cells just like an envelope. It provides nourishment to the embryonic cells and finally disintigrates. The *mesomeres* form the inner thick striated membrane known as *embryophore*. Beneath embryophore lies basement membrane.

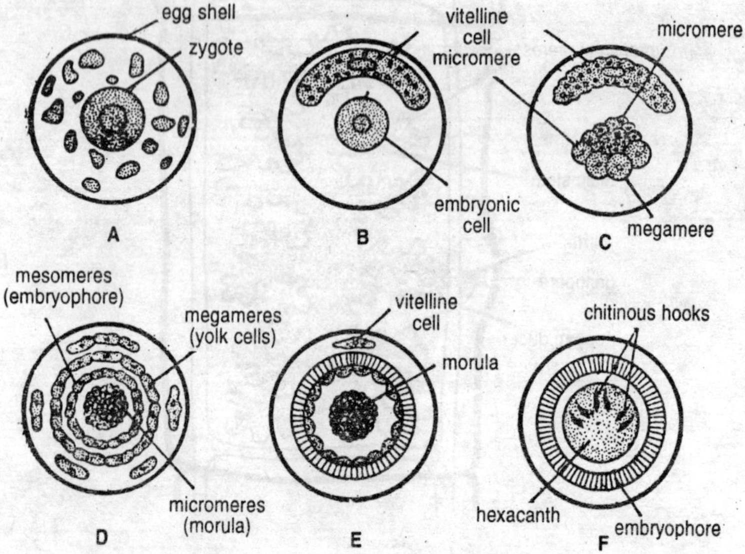

Fig. 6.11. *Taenia solium*. A--Zygote. B,C,D,E--Cleavage. F - Onchosphere.

The morula develops three pairs of chitinous *hooks*. These hooks are secreted by the differentiated cells known as *onchoblasts*. Now the six-hooked embryo is called *hexacanth*. It possesses a pair of large penetration glands and is surrounded by two membranes, the *hexacanth membranes*. The hexacanth together with all the membranes surrounding it is known as *onchosphere*. The onchosphere loses the shell.

This stage of development is found in the proglottids at the extreme posterior end of the body. The terminal proglottids are liberated from the body 4 or 5 at a time and are passed out quietly with the faeces of man.

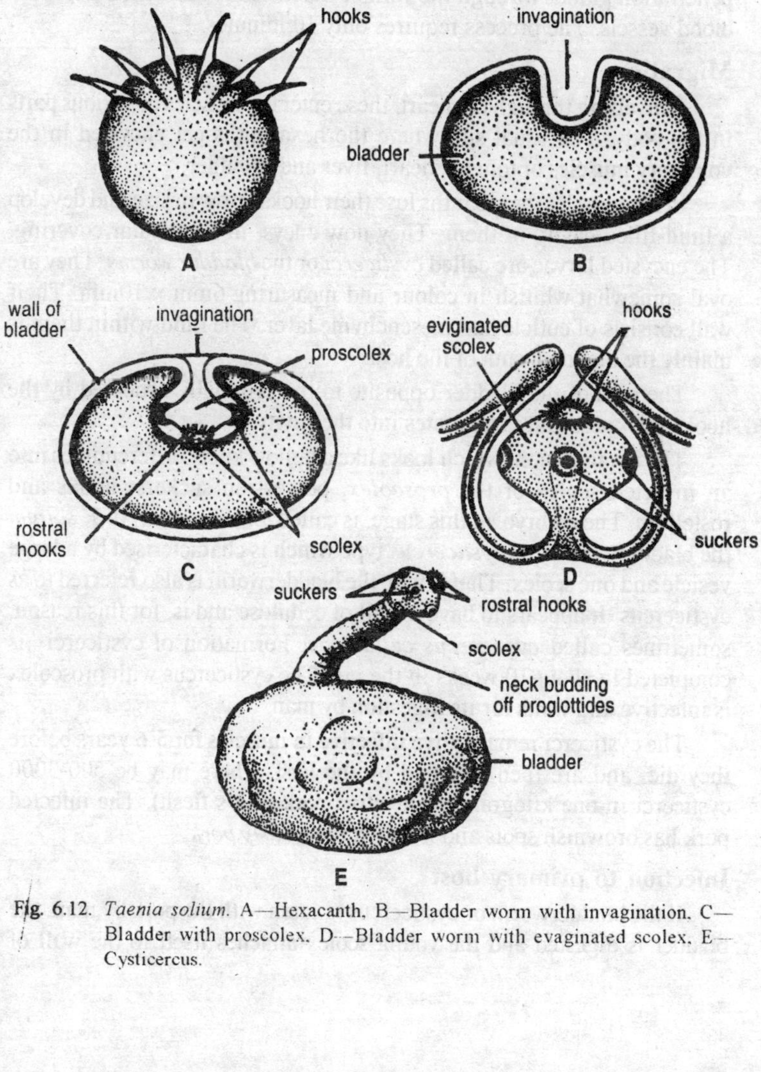

Fig. 6.12. *Taenia solium*. A—Hexacanth, B—Bladder worm with invagination, C—Bladder with proscolex, D—Bladder worm with evaginated scolex, E—Cysticercus.

The proglottids show slow movements of contraction and later on perish. *Embryos* remain active for a considerable time. They are capable of further development only if they are taken in by a secondary host, which is a pig.

Infection to secondary host

The pig is the secondary host which regularly feeds on the faeces of man. Dogs, monkeys and sheeps are also known to get infection.

Inside the alimentary canal of a pig, proglottids with thousands of egg's shells, nearly 30,000 to 40,000 are digested and *hexacanths* are set free. Later on, hexacanths bore their way with the help of hooks and penetration glands through the wall of the intestine and reach intestinal blood vessels. The process requires only 10 minutes.

Migration

Travelling through the heart, these enter the muscles of various parts in the body. The usual site where the hexacanths get encysted in the voluntary muscles of tongue, heart, liver and shoulder.

In the muscles, hexacanths lose their hooks, grow in size and develop a fluid-filled cavity in them. They now encyst in a cuticular covering. The encysted larvae are called *cysticerci* or the *bladder worms*. They are oval somewhat whitish in colour and measuring 6mm x 10mm. Their wall consists of cuticle and mesenchyme layer. The fluid within them is mainly the blood plasma of the host.

The side of the bladder opposite to the originally occupied by the hooks thickness and invaginates into the cavity.

The invagination, which looks like a hollow knob, differentiates into an inverted scolex or the *proscolex*, possession suckers, hooks and rostellum. The embryo, at this stage, is called *bladderworm*. In *T. solium* the bladderworm is of *cysticercus* type which is characterised by a large vesicle and one scolex. That is why the bladderworm is also referred to as cysticercus. It appears to have a wall of cellulose and is, for this reason, sometimes called *cysticercus cellulosae*. Formation of cysticerci is completed in about 10 weeks in the pig. The cysticercus with proscolex is infective and waits for its ingestion by man.

The cysticerci remain alive infective in muscles for 5-6 years before they die, and are then calcified by the host. There may be 500-3000 cysticerci in one kilogram of infective pork (pig's flesh). The infected pork has brownish spots and is called the *measly pork*.

Infection to primary host

If the bladderworm or cysticercus is eaten with the pork by man, the bladder is digested and the young scolex attaches itself to the wall of intestine by hooks and suckers. A region of proliferation of neck is formed, which develops proglottids of the adult tapeworm. While anchoring itself to the wall of small intestine in matures in two or three months.

Pathogenesis

Both adults as well as cysticercus stages are harmful:

1. Adult worms

The scolex of adult worm remains embedded in the mucous lining of alimentary canal for the attachment. The injury of musoca may lead to bacterial infection. There may be abdominal discomfort, diarrhoea, frequent hunger pangs, nervousness, insomnia, nausea and epileptic etc. These symptons are due the presence of adult worm are collectively known as *Taeniasis*.

2. Cysticercus larvae

Sometimes, man may accidentally ingest eggs or proglottids with contaminated food. Serious effects may result from *cycticercosis* in humans. Although they may be found in almost any organ, cysticerci are commonly located in striated muscles, liver, eye and the central nervous system, where they most often are in the meninges and ventricles of brain (*Marquez-Monter*, 1971). In an aberrant from of *T.solium,* apparently found only in the central nervous system, a cysticercus has daughter cysticerci budding from the bladder wall and forming an invasive, proliferating parasite. This type of metacestode is referred to as a *recemose cysticercus*. No viable scolices have been found in recemose cysticerci (*Beaver* and *Rolon*, 1981).

Living cysticerci usually elicit little inflammatory response. They damage the host by forming space - occupying leisons, that is, their growth puts pressure on the adjacent tissue and interferes with its function or causes necrosis due to reduced blood circulation. In the brain ventricle the organisms cause *hydrocephalus* because the cerebrospinal fluid cannot be properly servere inflammatory reactions.

Treatment

Niclosamide is the drug of choice for intestinal infections, it is highly effective and well tolerated. Carbon tetrachloride and oleoresin are also used as an anthelmintic. For removal of tapeworms, doses of 0.4gm each of *atabrin* and *acranil* or 1gm of *hexylresorcinol* with 2 gm of *acacia* may be given. Cysticercosis is best treated in human by surgical removal of cysts, when possible. If surgery is not possible meberidazol can be administered.

Epidemiology and control

Intestinal infections in humans are acquired by eating inadequately cooked pork. Pigs are infected by eating gravid proglottids passed with human faeces. Since a single proglottid may have hundreds of eggs, a pig may obtain a heavy infection throughout its body by eating only a few proglottids.

Following are the principle means of control :

1. Adequate cooking of food.
2. Freezing pork after slaughter and before it is marketed.
3. Sanitary disposal of human waste.
4. Meat inspection.
5. Prompt treatment of person found to be infected.
6. An educational program aimed at changing habits of personal hygiene so that person do not reinfect themselves or others they have contact with.

PARASITIC ADAPTATIONS IN *TAENIA*

Taenia is an endoparasite in the intestine of vertebrate hosts. It has undergone a number of structural modifications to adjust in its specific environment. The modifications due to parasitic mode of life can be discussed under following headings:

I. Structural
II. Physiological
III. Modifications in Life-cycle.

I. Structural modifications can be divided into :

(A) Change in external form
(B) Change in internal form

(A) Change in external form

1. Body is dorso-ventrally flattened and ribbon like.
2. Neck possess the power of forming new proglottids.
3. Scolex or the anterior end is modified into a hold-fast for attachment for this purposes scolex bears suckers and spines.

(B) Change in internal organs

1. Body wall is devoid of cellular epidermis and remains invested by a thick and resistant cuticle.
2. Alimentary canal is absent because the animal lives in intestine where digested food materials are always available.
3. Excrectory system consists of protonephridia and which helps in osmoregulation.

4. Respiration anaerobically.
5. Sense organs are absent except tangoreceptors.
6. Highly developed reproductive system.

II. Physiological adaptations

1. It can resist host's digestive enzymes through some unknown mechanism, probably by the secretion of anti-enzymes.
2. It maintains an equilibrium with the osmotic pressure of host.
3. It stimulates the mucous membrane of the intestine of host to secrete a large amount of mucous that cover for further protection against host's digestive enzymes.
4. Power of anaerobic respiration enables it to live in the oxygen free intestinal contents.
5. Long flattened body provides large surface area for absorption of digested food of the host that is why the mouth and alimentary canal are absent.

III. Modifications in Life-cycle

1. To cover the risk of finding individuals of different sexes in one host the organism is bisexual and moreover, each proglottid is bisexual and self-fertilization occurs.
2. Each proglottid produces 30-40 thousand eggs.
3. A secondary host i.e. pig is introduced to ensure its dispersal from one primary host to another.

Revision Questions

1. Describe the structure and functions of the body wall in a tape-worm.
2. Describe the reproductive organs of *Taenia.*
3. Describe the life-history of *Taenia solium.*
4. Describe how *Fasciola* and *Taenia* are differ from each other regarding their external features, excretory system and reproductive system.
5. What is parasitism. Give an account of parasitic adaptations of *Taenia.*
6. Describe anatomy of *Taenia solium* with special reference to its parasitic mode of life.
7. Draw labelled diagrams. No description is required : (a) Section of Body wall of *Taenia,* (b) Mature proglottid showing reproductive organs, (c) T.S. *Taenia* through ootype, (d) Cysticercus larva.
8. Write short notes on : (a) Hexacanth, (b) Apolysis, (c) Gravid proglottid, (d) Scolex, (e) Bladderworm, (f) Measly pork.

7

Echinococcus granulosus

Echinococcus granulosus or the 'hydatid worm', primarily parasetizes members of the canine family. The larval stage, the hydatid, is normally a parasite of sheep, but almost any mammal will serve as a host. It is potentially dangerous to human beings, in whom it may give rise to a fatal metastasizing, cancer-like tumor.

SYSTEMATIC POSITION

Phylum	-	Platyhelminthes
Class	-	Cestoda
Subclass	-	Eucestoda
Order	-	Cyclophyllidea
Genus	-	*Echinococcus*
Species	-	*granulosus*

History and Habitat

The hydatid disease, which is applied for any tumor or swelling of a cystic feature caused by *Echincoccus. Redi (*1684) and *Tyson* (1691) who had reported actual animal nature of hydatid. But it was first discovered by *Goeze* (1682). Next *Hartmann* (1695) who found the adult in the intestine of dog.

Thus *Echinococcus* is generally found in the small intestine of dogs, wolfs, foxes and other carnivorous mammals. But larvae of this parasite develop into hydratid cysts mainly in human beings and domestic individuals.

Distribution

Echinococcus is cosmopolitan in distribution, especially in sheep and cattle-grazing areas. It has been reported from Iceland, South Australia, Tasmania, New Zealand, southern South America, southern and northern Africa, southern and eastern Europe, Siberia, Mangolia, northern China, Japan, and the near East. In India, it has been reported from Punjab.

EXTERNAL FEATURES

Shape and Size

The adult of *E. granulosus* is typically cyclophyllidean (*Taenia*), but it differs from it in size. It measures from 2-8 mm. in length.

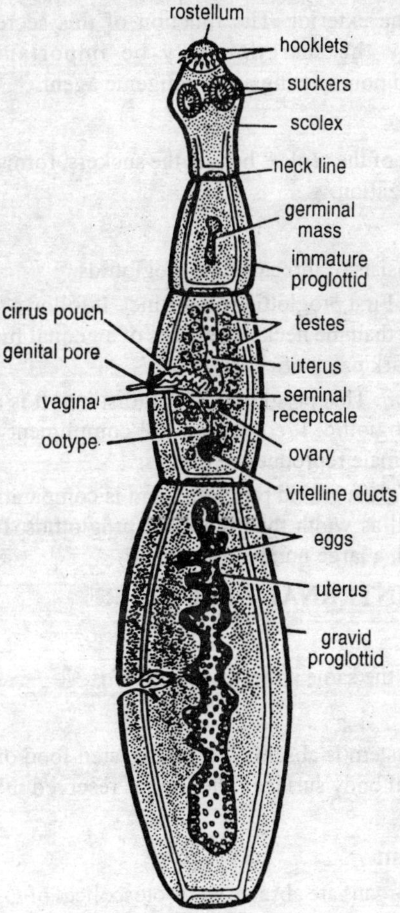

Fig. 7.1. *Echinococcus granulosus.* Entire.

Division of body. The body is divisible into (i) Scolex (ii) Neck (iii) Strobila.

(i) Scolex

The scolex is anteriormost part of body measuring 500μ in diameter.

The scolex bears four cup-like muscular structures, the *suckers* for the attachment to the intestine of host. Anteriorly situated, rectractile *rostellum* is armed with a double row of 20-80 hooks. *Smyth* (1963) has reported the presence of a group of cells lying just beneath the anterior tip of the *rostellum*. These cells not only contain droplets but actually secrete small viscid droplets to the exterior. The function of this secretion is still unknown probably the secretion may be important from the immunological stand point, acting as an antigenic agent.

(ii) Neck

The narrow part of the scolex, behind the suckers, forms the neck. It is the area of strobilization.

(iii) Strobila

The strobila consists of three or four proglottids.

(a) *First proglottid*. First proglottid is indistinct. It follows the neck and is slightly wider than the neck. It consists of a genital rudiment only in the form of dark patch of cells.

(b) *Second proglottid*. This proglottid is well distinct. It is much larger than the first progiottid. It contains a full compliment of taenioid-like male and female reproductive organs.

(c) *Third proglottid*. It is gravid proglottid and is comparatively bigger in length as well as width than the other proglottids. It contains a uterus filled with a large number of eggs.

INTERNAL STRUCTURE

Body wall

It is made up on the same pattern as in *Taenia*.

Digestive System

The digestive system is absent. The predigested food of the host is taken through general body surface. The food is reserved in the form of glycogen.

Respiratory System

The respiratory organs are absent. The protoscolices of *Echinococcus* carry out aerobic fermentation of carbohydrates. The protoscolices excrete pyruvic, acetic, succinic and lactic acids, and small amount of ethyl alcohol. In the absence of oxygen there is an increased production of succinate and no pyruvate is excreted. Carbon dioxide which is formed as byproduct is diffused out.

Excretory System

It is well developed containing longitudinal and transverse excretory canals with flame cells. Excretory system mainly help in regulating the amount of water in body, hence they are mainly osmoregulatory in function.

Nervous System

Well concentrated in the scolex.

REPRODUCTIVE SYSTEM

Echinococcus is hermaphrodite. The male and female reproductive organs are present in the second progiottid or mature progiottid.

Male Reproductive System

In the middle region of progiottid, encircling the other reproductive organs, are about 20-30 spherical structure known as *testes*. From each testis a fine duct arises known as *vas efferens*. In the postmedial portion of the proglottid the vasa efferentia join to the base of a *vas deferens*. Vas deferens is an extremely convoluted, sperm filled tube, which passes lateral to an enlarged, coiled, marginal *cirrus* or *penis*. The cirrus lies in a bulbous *cirrus pouch* or *sac*. The cirrus, during copulation is protruded through a lateral *genital atrium* and *genital pore*.

Female Reproductive System

Opening into the genital atrium closely behind the cirrus is a female copulatory tube, the *vagina*. The vaginal opening is guarded by a muscular sphincter. Beyond thc sphincter, the vagina passes downwards and enlarges at its inner end to form a sperm filled sac, the *seminal recepticle*. On the posterior side of the proglottid there is a bilobed *ovary* with a narrow, *central isthmus* from which an *oviduct* arises posteriorly. The oviduct joins a duct from the seminal recepticle, and this combined fertilization duct continuous posteriorly into a median compact *ootype*. The ootype is surrounded by unicelluar *Mehlis' glands*. In the posterior most region there lies a bilobed *vitelline gland*. The two lobes of vitelline glands are oriented dorsoventrally. From the vitelline gland a median *vitelline duct* arises that join the ootype. From the ootype a tube-like *uterus* arises which extends to a point near the anterior border of the proglottid.

Eggs

Eggs, resembling those of taenoid, are monoperculate and measuring about 30 × 38 μ. These are *ectolecithal* i.e. the zygote is surrounded by the yolk cells provided by vitelline gland and shall material. After an egg-shall is laid down the eggs pass from the ootype into the uterus. When the uterus of a gravid progiottid bursts, the eggs are liberated and passed out alongwith faeces of dogs. These eggs gain enterence into the intermediate host (man or other mammal) either through water or in forage. Each egg contains a fully developed *onchosphere*. Human infection develops from too intimate an association with dogs. Infection of children is commonly established when dogs are permitted to lick their mouths. Eggs can be ingested from contaminated fingers.

The eggs, upon being swallowed by a mamalian intermidiate host, hatch in duodenum as a typical six-hooked onchosphere. The hatching occurs in slight alkaline medium. Activation of embryos take place in either humen bile or intestinal juice of ruminants, they are not activated in bile or intestinal juicse of dogs and cats. The escaping onchospheres penerate the intestinal wall, enter the mesenteric venules, and become lodged in the capillary beds of various organs. The hooks are now shed. After three hours of egestion, the embryo reachs its final place of development. Among mouse, deer etc. they reach lungs where as in man and other mammals, they reach in the liver, lungs, brain, kidneys, spleen etc. but liver being the most common site. Implantation and growth occur here, onchosphere now develops into *hydatid cyst.* The cyst formation is a slow process, with the host laying down an envelope of connective tissue around the parasite. In about 4-5 months, the cyst reachs approximately 10 mm. in diameter. It goes on increasing in size and eventually becomes as big as the orange or even sometimes a football.

The various types of larval cysts, formed in the life-history of *Echinococcus* are:

1. Unilocular hydatid cyst

The hydatid cyst is composed of three layers. The inner most layer is called *endocyst.* It is thin syncytium forming the germinal layer. The middle layer is called *ectocyst*. It is made up of cuticle which is amorphous and laminated. The outer layer is *pericyst.* It is composed of connective tissue and fibres secreted by the host.

The cyst produces daughter cysts upon invagination of its wall. These daughter cysts project into the lumen of mother cyst. The cavities are filled with a sterile fluid known as *hydatid fluid.* Within some daughter cysts, tertiary or grand-daughter cysts are formed by invagination. These are endogenous or exogenous cysts or brood capsules.

The endogenous cysts fall free into the hydatid fluid of mother cyst while the exogenous cysts, after detaching, migrate to other part of body, where they continue to grow further. The daughter cysts have a fibrous cyst wall internally lined with endocyst.

Rounded and hollow brood capsules are also continuously budded off from the endocyst of mother cyst. These capsules remain attach for some times by means of a narrow pedicle.

From the wall of these brood capsules, minute scolices are formed. Scolex formation may occur from the invagination of walls either of daughter or grand-daughter cyst.

The anterior end is inturned to form a central canal into which the four suckers open and into the expanded base of which the hooklets project. The body of each scolex is filled with mesenchymal cells aggregated to form a subcuticular layer and to form a compact cup-shaped mass below the hooklets. The syncytium of the endocyst retains its embryonic potentialities, and new brood capsules may be produced indefinitely. Hydatids thus may expand to enormous size and contains nearly 50 litres of fluid as well as thousands of scolices. The pressure so produce may severely damage adjacent host organs.

When the infective brood capsule is ingested by the final host, each scolex quickly evaginate and attach to the duodenal mucosa. For the first four to five days in the host they undergo little growth. By the 14th day

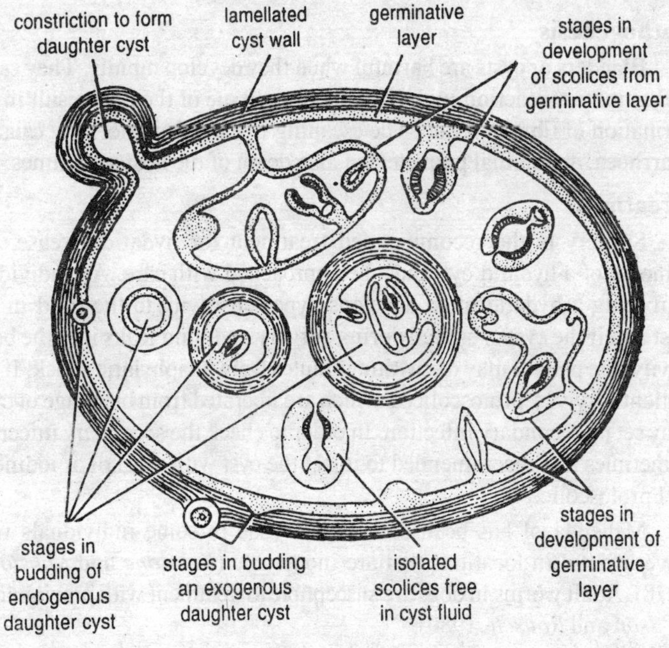

Fig. 7.2. *Echinococcus granulosus.* Hydatid cyst.

the scolex has grown in size, and first proglottid has appeared. The genital organs enlarge and a second proglottid appear by the 18th day. Growth continues and fully developed shelled embryos are formed by about the 35th day, and the life cycle begins again.

2. Multilocular cyst

It is characterized by many small bladders embedded in a common enclosing membrane. These are found in cattles. It is considered to be the result of differential reaction inside the host.

3. Alveolar hydatid in man

It is characterized by a malignant type of growth with jelly-filled proliferating vesicles embedded in a common dense stroma. The very thin laminated membrane does not restrict the germinal membrane that grows out into surrounding host tissues. It is commonly sterile and it undergoes degeneration and clacification.

4. Multivesicular

It is characterized by many adjoining and connected bladders, each having its own separate germinal membrane.

Pathogenesis

Bladders or cysts are harmful when they develop rapidly. They cause inflammatory reaction in the surrounding tissue of the host result in the formation of fibrous tissue. The escaping hydatic fluid not only causing diarrhoea, abdominal pain etc. but also death of the host sometimes.

Treatment

Surgery is the recommended treatment for hydatid disease, but removal of a hydatid cyst must be approached with care. An individual harboring a hydatid often develops hypersensitivity to the fluid in the cyst and if the cyst is broken during surgery and fluid leaks into the body cavity, the patient may die within minutes from anaphylatic shock. If the patient survives, protoscolices, which are liberated from breakage of cyst, may set up secondary infection. In order to cheek the secondary infection sometimes it is recommended to inject the cyst with alcohol or iodine to kill protoscolices.

Mebendazol has been successfully used in some individuals who have hydatid in locations that are inoperable (*Keystone* and *Murdoch*, 1978). Adult worms in dogs are susceptible to treatment with praziquantel (*Wilson* and *Rausch*, 1980).

Control

The followings are the elements of a control programme for hydatid disease:

1. Treatment of infected dogs with an effective anthelmintic or to kill them.
2. Keeping the viscera of domestic animals from being eaten by dogs.
3. Children should not be allowed to play dogs frequently.
4. Avoiding play with dogs and hands should be thoroughly washed before eating.

Revision Questions

1. Describe the structure and reproductive system in *Echinococcus.*
2. Describe various types of larval cysts in life-cycle of *Echinococcus.*
3. Write short notes on : (a) Scolex, (b) Onchospheres, (c) Hydatid cysts.

8

Aschelminthes: Characters & Classification

There is variety of opinions among zoologists about the systematic position of nematodes. Some authors (*Gegenbaur* etc.) have designated a separate Phylum Nemathelminthes, which includes or does not include Acanthocephala and Nematomorpha. Recently, nematodes have been grouped in a class Nematoda in the Phylum Aschelminthes. The term Aschelminthes is coined by *Grobben*.

Characters

1. Round worms have elongated, cylindrical and vermiform body with tapering ends.
2. Body is unsegmented but may be wrinkled. It is not distinguished into regions.
3. Body unpigmented being either white or with a yellowish tinge.
4. Caudal end of the body generally straight in female but coiled in males and the males are shorter than females.
5. Anterior cephalization is not prominent, therefore, the body is without any definite regions.
6. Mouth terminal surrounded by lips. In *Strongyloides* the lips are modified into teeth known as leaf-crown.
7. Amphids and papillae are the main sensory organs and are of great taxonomic value in case of free-living forms.
8. Marine nematodes are annulated due to the presence of transverse striations or longitudinal striations which are of common occurence.
9. Body is covered by rough resistant cuticle, having bristles, spines, warts and papillae etc. Sometimes cuticle modified as cephalic, cervical and lateral cuticle and in the form of vesicle.
10. Caudal end with a pouch of cuticular nature known as *phasmids*, very common in parasitic forms.
11. Cuticle in the strongyloides is modified into an umbrella-like form known as *bursa* which is generally supported by muscular rays.

12. The layers of body wall generally made up of cuticle, sub-cuticle or epidermis and muscle layer. The cuticle is generally of keratin which allows only glucose and urea etc.
13. Epidermis is syncytial where they are usually divided in four sections in the four longitudinal chords. One dorsal, one ventral and 2 lateral in position.
14. Muscles consist of longitudinal muscle fibres with variable arrangements which is generally *holomyarian, meromyarian* and *polymyarian* etc. The muscles are of great taxonomic value.
15. The body cavity is pseudocoel.
16. Digestive tract well developed generally made up of mouth, buccal cavity, pharynx or oesophagus are of various types, intestine and anus.
17. Nervous system consists of a nerve-ring encircling the oesophagus. From it, nerves are given out anteriorly and posteriorly.
18. Protonephridia absent but the excretory system is made up of canals or gland-like organs.
19. Sexes are separate. Testis and ovaries are tubular and coiled. Usually there is a single testis. The ducts from the testis open into the cloaca and the cloaca is associated with accessory structures such as circular spicules. Ovaries, oviduct and uteri are double.
20. Aschelminthes are ovo-viviparous, oviparous or viviparous.
21. Life-cycle complicated may be with or without intermediate host.
22. Larval stage four. Third stage of larva is infective.

Classification

Phylum Aschelminthes is divisible into one class-Nematoda.

CLASS-NEMATODA

1. Body is slender, cylindrical, tapering towards both the ends.
2. Body is covered with cuticle.
3. Intestine is well formed.
4. Body cavity is not lined with epithelium.
5. Cloaca is absent in female.
6. Male and female reproductive organs are well developed.

The Classification of Nematoda is not stable as they have been studied mainly independently. The scheme given below will be found simple and convenient.

Class Nematoda is divided into two sub-classes :

Sub-class I-Aphasmidea

Sub-class II-Phasmidea

Sub-class I-Aphasmidea

1. Phasmids or caudal sensory organs are absent.
2. Amphids or anterior sensory organs many.
3. Excretory organs either reduced or absent.
4. Coelomocytes and mesenterial tissue are well developed.
5. Caudal alae in females are absent, free living.

The Sub-class is divided into following orders:

Order (1) Chromadoroidea

1. Oesophagus is not elongated but differentiated into three regions.
2. Cuticle smooth or ringed, ornamented and sometimes with bristles.
3. Amphids spiral.
4. Mainly marine and free-living.

Examples : *Paracytholaimus, Paracanthonchus.*

Order (2) Enploidea

1. Mainly marine and free-living.
2. Six sensory papillae on the lips.
3. Ten to twelve sensory bristles arranged in one or two circlets.
4. Cyathiform (goblet-shaped) amphids.
5. A pair of cephalic slits.
6. Size small to moderate.

Examples: *Pelagonema, Enoplus, Anticoma.*

Order (3) Dorylaimoidea

1. Found in fresh-water and soil.
2. Cuticle smooth without bristles.
3. Head with an inner circlet of six and an outer circlet of ten sensory papillae.
4. Odontostylet, the piercing apparatus derived from a modified tooth in the oral region.
5. Amphids cyathiform.
6. Posterior part enlarged.

Examples: *Dorylaimus, Xiphinema.*

Order (4) Mermithoidea

1. Anterior end with 16 labial papillae.

2. Sensory bristle absent.
3. Cuticle is smooth.
4. Amphids either rudimentary or cyathiform.
5. Oesophagus opening in blind intestine.
6. Larva parasites in invertebrates whereas adults are free-living

Examples: *Mermis*, *Paramermis*.

Order (5) Trichuroidea

1. Intestinal muscles, caudal gland or suckers are absent.
2. Mouth without lips.
3. Pharynx very long and narrow.
4. Females with one ovary.
5. Males with one or no copulatory spicule.

Examples: *Trichinella*, *Trichuris*.

Order (6) Desmoscalecoidea

1. Small sized with heavily ringed cuticle.
2. The head is armoured. Anterior end with four sensory bristles.
3. Amphids are crescent shaped or pump-shaped.
4. They are marine and free-living.

Examples: *Tricoma*, *Desmoscolex*.

Order (7) Monohysteroidea

1. Small forms with smooth or ringed cuticle having bristles.
2. Amphid circular.
3. Anterior end with 4, 6, 8, or many sensory bristles.
4. Free living forms, few are marine, some are fresh water and others terrestrial.

Examples: *Monohystera*, *Siphonolaimus*.

Sub-class II- Phasmidea

1. With phasmids and pore like amphids.
2. Excretory organs with paired lateral canals.
3. Coelomocytes 4-6.
4. Caudal or hypodermal glands absent.
5. Reproductive system well developed.

The sub-class has been divided into following orders:

Order (1) Rhabditoidea

1. Mouth without lips and with 6, 12, or 18 papillae.

2. Cuticle is ringed or smooth.
3. Pharynx, long with posterior bulb.
4. Single ovary in females and males with copulatory spicules.
5. Mostly parasite on birds and mammals.

   Examples: *Rhabditis*, *Diplogaster*.

Order (2) Oxyuroidea

1. Pin-shaped small or medium sized.
2. Cervical papillae absent.
3. Pharynx with posterior bulb.
4. Female with slender tail.
5. Males with one or two copulatory vesicles and sometimes with alae forming a bursa.
6. Parasites in invertebrates and vertebrates.

   Examples: *Enterobius* or *Oxyuris*.

Order (3) Ascaroidea

1. Large parasitic.
2. Mouth surrounded by three rerely six lips.
3. Cervical papillae present.
4. Pharynx with or without valve.
5. Female short tail.
6. Male with ventrally coiled tail and with two equal copulatory spicules and usually without alae.

   Example: *Ascaris*.

Order (4) Strongyloidea

1. Mouth without distinct lips but has leaf crowns.
2. Pharynx is without bulb.
3. Males are provided with a distinct caudal bursa which is supported with rays.
4. Females with an ovejactor.

   Examples: *Ancylostoma*, *Necator*, *Strongylus*.

Order (5) Dracunculoidea

1. There are no definite lips or buccal capsule.
2. Mouth is surrounded by ring of papillae.
3. Pharyngeal bulb is absent. Pharynx is muscular anterior but glandular posteriorly.
4. Copulatory bursa is absent in males.

5. They are parasite of vertebrates.

   Examples: *Dracunculus*, *Micropleura*.

Order (6) Spiruroidea

1. Mouth with two lateral lips, sometimes 4 or 6.
2. Pharynx with bulb with anterior muscular and posterior glandular part.
3. Males without bursa but have spirally coiled tail.
4. Parasitic forms.

   Examples: *Gnathostoma*, *Spiroxys*.

Order (7) Rhabdiasoidea

1. Cuticle smooth.
2. Definite pharyngeal bulb is absent.
3. Parasitic stage is hermaphrodite or parthenogenetic free-living stage may develop into males and females.

   Example: *Rhabdias*.

Order (8) Filariodea

1. Slender filiform worms.
2. Mouth simple, lips are absent.
3. Females larger having an anterior vulva and a long tubular vagina.
4. Males with two unequal spicules. Alae present or absent.
5. Found in blood or lymphatic system. Transmitted by blood sucking insects.

   Examples: *Wuchereria, Microfilaria, Loa.*

TRICHURIS

Phylum	-	Aschelminthes
Class	-	Nematoda
Sub-class	-	Aphasmidea
Order	-	Trichuroidea
Genus	-	*Trichuris*

1. It is commonly called as *whip-worm.* It is an endoparasite found in the caecum and large intestine of man.
2. It has world wide distribution but common in tropical countries.
3. Female is slightly larger than males.
4. The anterior part of the body having pharynx is usually more slender and this portion is burried in the mucosa of host's alimentary canal.
5. Mouth is simple without lips.

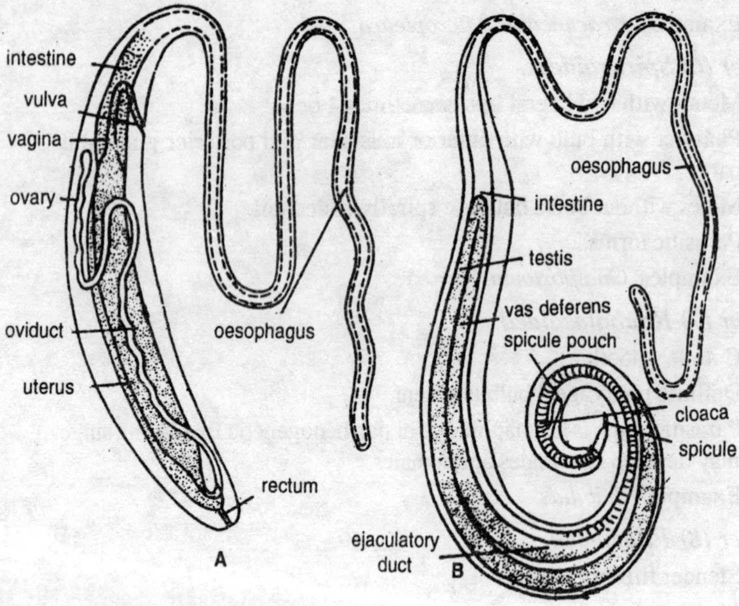

Fig. 8.1. *Trichuris.*

6. It feeds on blood and epithelial cells.
7. Life-cycle is simple with no intermediate host.
8. Development of eggs take place in moist soil. Embryos develop in about three weeks.
9. Contaminated food and water responsible for infection in man by swallowing eggs with juveniles.
10. Liberated juveniles grow into adults within a month.
11. Excretory system is absent.
12. Heavy infections is responsible for diarrhoea and anaemia.

TRICHINELLA

The systematic position is the same as that of *Trichuris.*

1. It is a parasite of man, pig, rat and other vertebrates. It is commonly known as *trichinia worm*.
2. It is commonly found in Europe, United States and Arctic regions.
3. Sexes are separate, males are about 1.5 mm. in length and the female are 3 mm. in length.
4. In both the sexes, the mouth is without lips.
5. Pharynx is long, slender and non muscular and is embedded in stichosome.

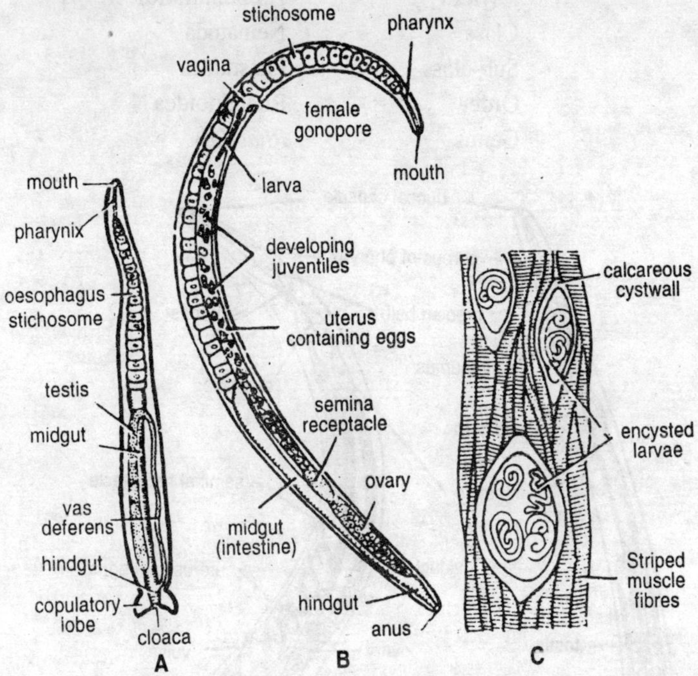

Fig. 8.2. *Trichinella.* A- Adult male, B-Adult female, C- Encysted larvae in stripe (voluntary) muscle of pig.

6. The oesophagus is tubular, well developed and occupies 1/3 of the total length of body.
7. The female worms which are viviparous enter the lymphatic and reach the muscular system where they secrete a cyst around them. This is called encysted sexless stage.
8. The encysted sexless trichinae are ingested by man along with pork-flesh, where the cyst wall is dissolved by the action of digestive juice and the reproductive organs start developing.
9. The developing worm begins to migrate into the muscles and produces the symptoms of a dreaded disease call *Trichinosis.*
0. Patients feel gastro-intestinal pain, and weakness with or without fever.

RHABDITIS

Phylum	-	Aschelminthes
Class	-	Nematoda
Sub-class	-	Phasmidea
Order	-	Rhabditoidea
Genus	-	*Rhabditis*

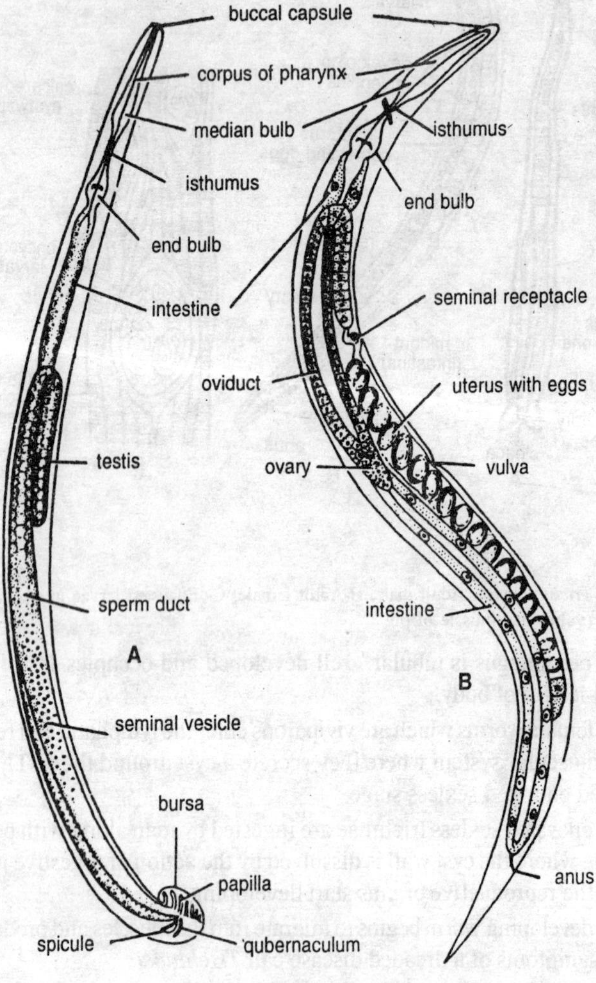

Fig. 8.3. *Rhabditis. A-male, B-female*

1. Genus *Rhabditis* includes numerous free-living, semiparasitic forms. They are commonly found in organic matter, faeces of man. They feed on the fluid formed by bacterial decying of organic matters.
2. The pharynx has two bulbs with an isthmus joining them.
3. The excretory system is H-shaped.
4. The reproductive system is J-shaped.
5. The large female has a pointed tail and didelphic sex-organs i.e. having paired ovaries, oviducts, seminal recapticles and uteri.
6. The male has a flattened copulaltory bursa, two equal spicules and monorchic sex organs i.e. single testis.
7. Life-cycle is simple and direct. The juveniles adhere to the bodies of dung beetles for transport to fresh dung.

ENTEROBIUS VERMICULARIS

Phylum	-	Aschelminthes
Class	-	Nematoda
Sub-class	-	Phasmidea
Order	-	Oxyuroidea
Genus	-	*Enterobius*
Species	-	*vermicularis*

1. It is commonly called *'Pinworm'* or *'seatworm'*.
2. It has world-wide distribution, but specially found in Europe, India, Africa, Canada, Chile and U.S.A.
3. Adult worms live in the caecum, appendix and adjacent parts of intestine of man.
4. Sexes are separate. The male measures 2-5 mm. in length and the female measures 8-13 mm. in length.
5. Anterior end is provided with three lips and a pair of cephalic expansion.
6. Male has ventrally curved tail with a single spicule and a pair of cuticular expansions known as *caudal alae*. It is monorchic.
7. Female has a straight tail which becomes long and pointed. It is didelphic. Vulva is anterior.
8. Life-cycle is simple there is no intermediate host. Gravid females migrate to anus causing itching sensation. They crawl and deposit eggs on the perineal skin. Due to scratching, gravid females rupture and thus small eggs are liberated. The re-infection in children takes by putting fingers in mouth. Eggs hatch. Juveniles inhabit small

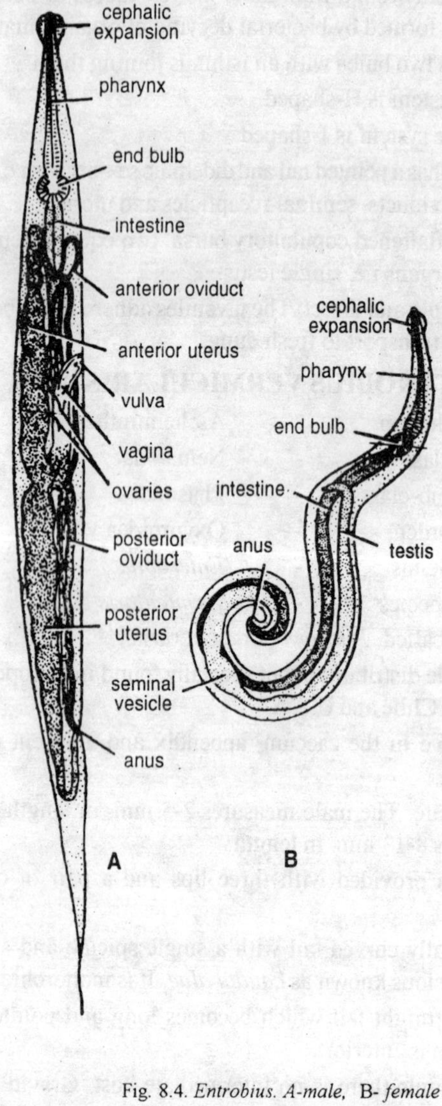

Fig. 8.4. *Entrobius. A-male,* B- *female*

intestine. They moult twice and turns into adults which reach their normal habitat. The infection is called *Enterobiasis*. Symptoms are inflammation of caecal mucosa, abdominal pain, insomnia, hysteria, restlessness, loss of appetite and sometimes appendicitis.

LOA LOA

Phylum	-	Aschelminthes
Class	-	nematoda
Sub-class	-	Phasmidea
Order	-	Filarioidea
Genus	-	*Loa*
Species	-	*loa*

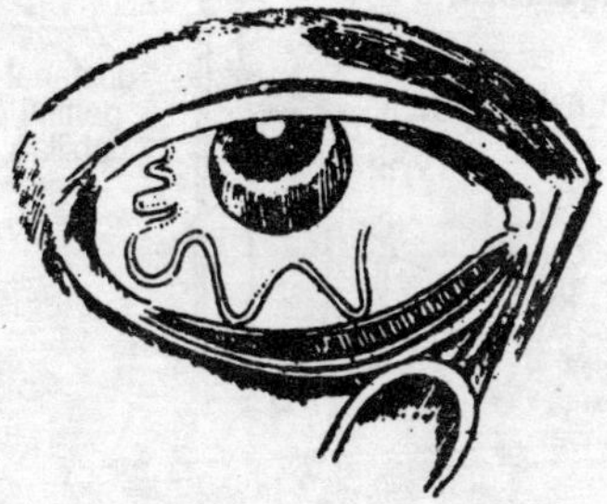

Fig. 8.5. *Loa Loa* in the cornea of man.

1. It is commonly known as *African eyeworm*, chiefly found in Africa.
2. They live in the subdermal connective tissue of man in West Africa.
3. The male and female measure 20-35 mm. and 20-70 mm. in length respectively.
4. Body is covered by numerous warts.
5. The intermediate hosts are mango flies, *Chrysops*. The larva moult and develop upto infective stage in fly. These are transmitted in blood of man through proboscis.
6. The adult worm moves about in the body in the subdermal tissue and sometimes passes across the eye ball causing intense itching and swelling known as '*Calabar Swellings*'.
7. These are injurious and fatal when they penetrate brain and spinal cord.

DRACUNCULUS

Phylum	-	Aschelminthes
Class	-	Nematoda
Sub-class	-	Phasmidea
Order	-	Dracunculoidea
Genus	-	*Dracunculus*
Species	-	*medinensis*

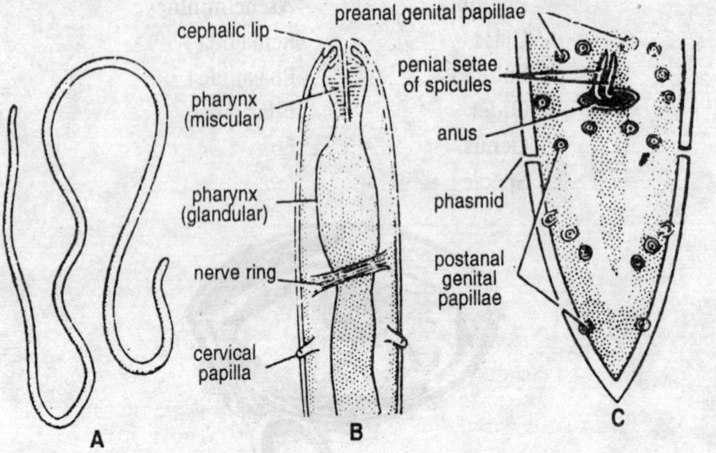

Fig. 8.6. *Dracunculus medinensis.* A- Entire, B-Anterior end of male. C-Posterior end of male ion ventral view.

1. It is commonly called '*Guinea Worm*'. It is found in tropical countries like India, Africa, South America and West Indies.
2. It lives in the deeper subcutaneous tissues of man.
3. The anterior end is blunt while the posterior end is pointed.
4. The adult male measures 4 cm. in length. Its hind end is curved and bears 10 pairs of genital papillae and a pair of pineal spicules.
5. Adult female measures 70-120 cm. in length. Its hind end is straight and bears a spine-like process.
6. The oesophagus is distinguished into an anterior narrow muscular region and a posterior broader glandular region.
7. The female reproductive system is didelphic with opposed uteri.
8. Females are viviparous. The alimentary canal and vulva are atrophied in mature females.
9. Mature female migrates to the superficial layers of the skin in those parts of the body which are likely to come in contact with cold water. Toxic substances are produced under the skin which results in the formation of a blister, leading to an ulcer, which breaks on coming in contact with water. On coming in contact with water, the worm ruptures, releasing larvae in water. The larvae swims about but do not develop further unless swallowed by female *Cyclops*, which serves as intermediate host. Here the female larvae moult twice and become

infective. Infection occurs when *Cyclops* are swallowed by a new host alongwith drinking water.

10. It causes *guinea worm disease* characterised by *asthma, nausea, vomiting, diarrhoea* etc.

Revision Questions

1. Ennumerate the characteristics of phylum Aschelminthes. Name its classes and discuss any of these.
2. Classify nematoda giving diagnostic characters and examples of each group.
3. Give an account of characters of Aschelminthes. Classify it up to order giving examples.
4. Discuss the relationship of nematodes with other animals.
5. Write short notes on : (a) *Entrobius vermicularis,* (b) *Wuchereria bancrofti*, (c) *Loa,* (d) *Trichinella spiralis.*

9

Ascaris lumbricoides

Much is known about the morphology and life-cycle of Nematodes but out of them *Ascaris* is most known. *A. lumbricoides* is the most common nematode parasite found in the human intestine. Being of large size and found abundantly, *Ascaris* is studied as the main example of the nematodes.

SYSTEMATIC POSITION

Phylum	-	Aschelminthes
Class	-	Nematoda
Subclass	-	Phasmidea
Order	-	Ascaroidea
Genus	-	*Ascaris*
Species	-	*lumbricoides*

Habits and Habitat

A. lumbricoides, a large nematode parasite of man, is world-wide in distribution. It occurs more commonly in the small intestine of children than in adults because of the children's custom of playing in contaminated, egg-laden soil adjacent to dwellings. As many as 1000-1500 adult worms may be present in a single host. Because of their strong, pointed and cuticulate bodies and their habit of wandering, they may perforate the gut wall and cause peritonitis or they may enter the bile or pancreatic ducts or the vermiform appendix and interfere with normal digestive functions.

EXTERNAL FEATURES

Shape and Size

Both the sexes are elongated, cylindrical and pointed at both the ends. Males are 15-31 cm. long and 2-4 mm. in diameters, while the females are 20-35 cm. long and 4-6 mm. in width. The males are further distinguished by a ventrally curved posterior end, from the slit-like opening of which often protrude two copulatory spicules.

Colouration

In the living condition, the colour of the animal is milky-white or reddish-yellow. There is a characteristic lustre on the body.

Longitudinal lines

There are four longitudinal lines or streaks, running throughout the length. One is mid-dorsal, one mid-vertral and two lateral. The dorsal and ventral lines are white in colour, the lateral streaks are thicker and broader than dorsal and ventral streaks and are brown in colour.

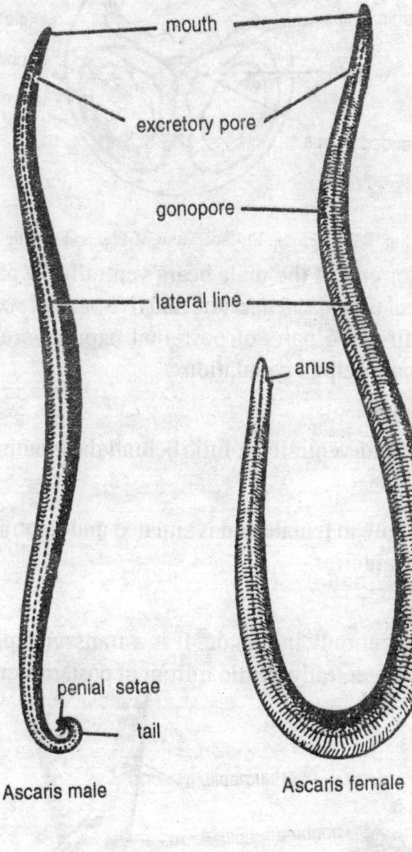

Fig. 9.1. *Ascaris lumbricoides.*

Mouth

A triangular mouth is present at the anterior end of the body. It is surrounded by three denticulated lips. One of these is mid-dorsal and other two are latero-ventral in position. The mid-dosal lip is almost elliptical and has a fork fleshy core in the centre. A circle of four sensory papillae is found on it. The remaining two lips also have a forked core and two rows of *sensory papillae*. Each sensory structure is called *amphids*. The sensory papillae of nematodes are very useful in the identification of the species and assigning their position in the classfication. The sensory papillae are *tangoreceptors*. The amphids are *chaemoreceptors*.

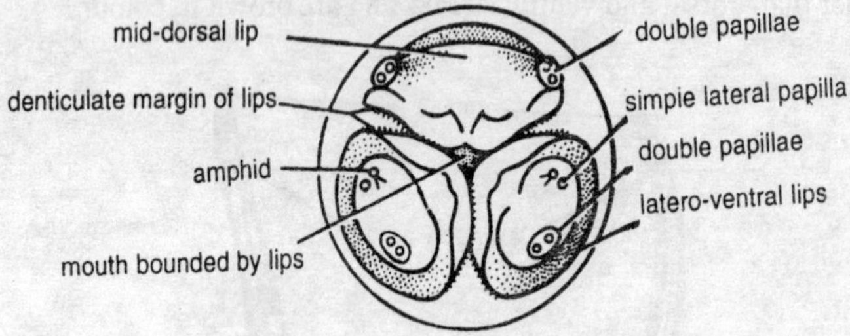

Fig. 9.2. *Ascaris*. En-face view of lips and mouth.

The posterior end of the male bears ventrally 50 pairs of *pre-anal papillae* in front of the cloacal aperture and five pairs of *post-anal papillae* behind it. The first two pairs of post-anal papillae are double. These papillae of the male help in copulation.

Excretory pore

It is situated mid-ventrally, a little behind the mouth.

Genital pore

It is present only in female and is situated mid-ventrally at about 1/3 distance from the anterior.

Anus

It is also present only in female. It is a transverse pore guarded by thick lips, lies mid-ventrally a little infront of posterior end.

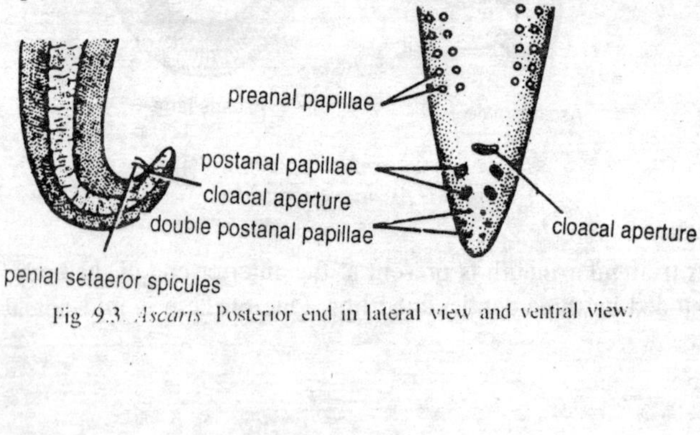

Fig 9.3. *Ascaris*. Posterior end in lateral view and ventral view.

Cloacal aperture

It is present only in male. It is situated mid-ventrally in front of the curved posterior end. Two chitinous *penial setae* or *penial spicules* are protruding through cloacal aperture.

BODY WALL

The body wall is made up of three layers.

1. Cuticle
2. Epidermis
3. Muscle layer

1. Cuticle

The cuticle forms a thin transparent, elastic but resistant covering around the body. It protects the body against the action of digestive juices. It is secreted by underlying epidermis. Histologically, the cuticle is formed of several layers and can be distinguished into (i) Cortical layer, (ii) Matrix, (iii) Fibrous layer and (iv) Basement membrane.

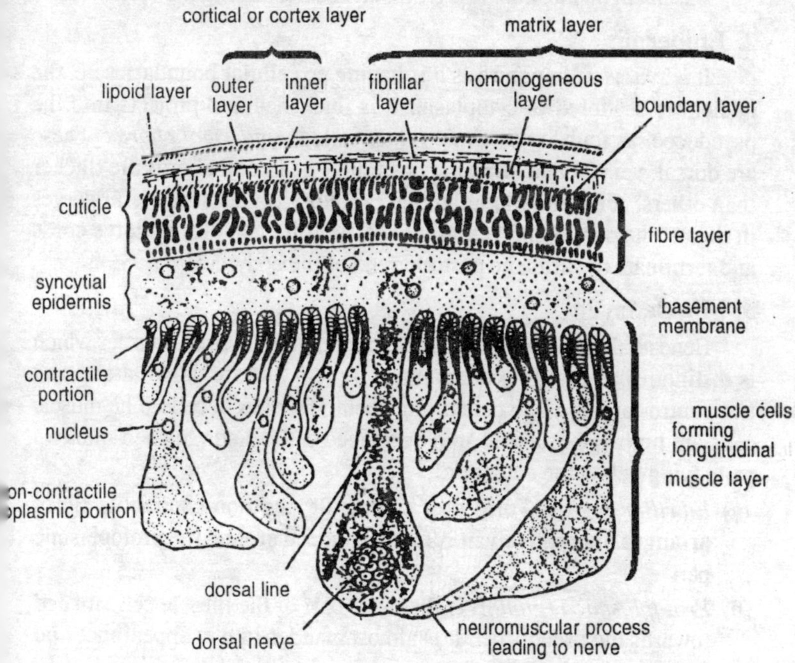

Fig. 9.4. *Ascaris.* Body wall in section.

(i) *Cortical layer.* It is made up of dense keratin. It protects the animal from the digestive juices of host.

(ii) *Matrix.* It is an elastic layer containing sulphur rich matricin.

(iii) *Fibrous layer.* It is made up of collagen fibres that cross each other obliquely.

(iv) *Basement membrane.* It forms the inner limit of cuticle.

Under electron-microscope the cuticle is made up of following layers:

1. *Lipoid layer.* It is in the form of a thin osmophilic membrane, about 1000 A° thick.
2. *Cortical or cortex layer.* It consists of (a) an *outer cortical layer* which lies as discontinuous strips or rings around the body, and (b) an *inner cortical layer.*
3. *Matrix layer.* It consists of (a) an outer *fibrillar layer* tranversed by branching canals, (b) a *homogeneous layer* with several radial striations, and (c) a *boundary layer* resembling the outer cortical layer.
4. *Fibre layer.* It consists of collagen fibres arranged in three strata.
5. *Basement membrane.* It is a thin layer surrounding the epidermis.

2. Epidermis

It is *syncytial* in nature as it is having no cellular boundaries i.e. the nuclei are scattered in cytoplasm. It is thickened and projects into the pseudocoel as four longitudinal ridges called, *epidermal chords*. These are dorsal, ventral and lateral in position. The lateral chords are thicker than others, containing longitudinal excretory canals and lateral nerves in them. The dorsal and ventral chords are thin containing nerve cords and terminate towards the posterior end of the body.

3. Muscle layer

Beneath the epidermis is a single layer of longitudinal muscles which is distinguished into four quadrants. Of them two are dorso-lateral and two ventro-lateral. Each containing about 150 muscle cells. The muscle cells are highly specialised spindle-shaped cells, distinguished into two parts lying below the ectoderm.

(a) *Fibrillar region.* Formed of contractile and non-contractile fibres arranged alternatively, it is found wrapped around the protoplasmic part.

(b) *Protoplasmic region.* It is the inner part of the muscle cell situated towards the pseudocoel. It is almost bladder-like in appearance and consists of a single nucleus.

There is a complete absence of circular muscles in nematodes.

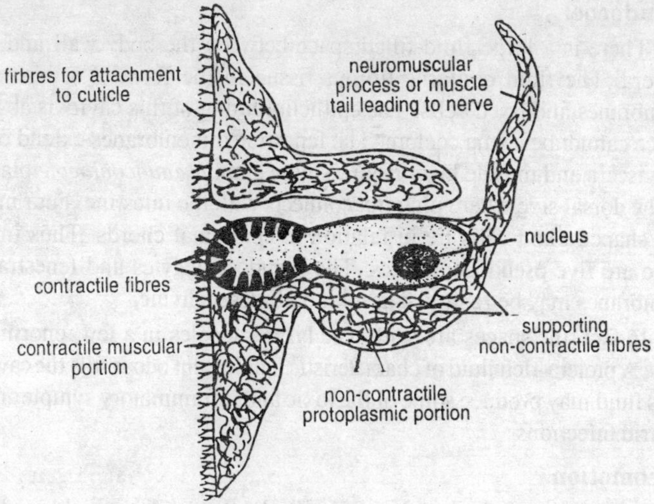

Fig. 9.5. *Ascaris.* A muscle cell in T.S.

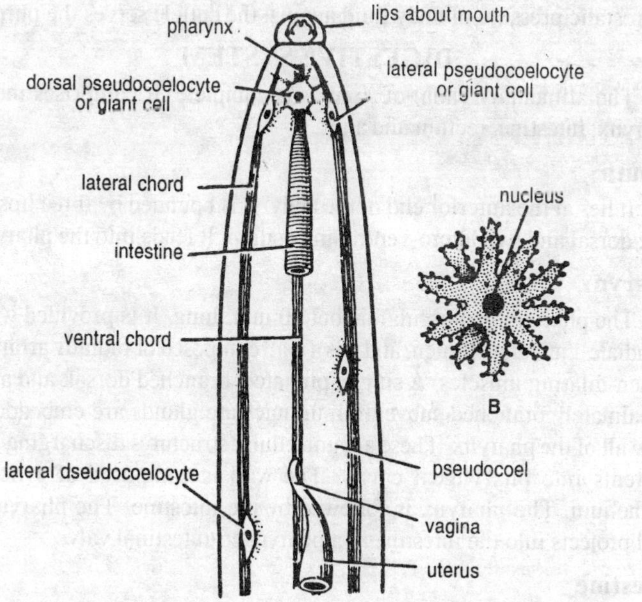

Fig. 9.6. *Ascaris.* A—Portion of giant cells in pseudocoel. B—Single giant cell.

Pseudocoel

There is a large fluid-filled space between the body wall and the viscera. The fluid contains fibrous tissue in the form of fenestrated membranes and fixed cells. The epithelial lining of this cavity is absent hence cannot be a true coelom. The fenestrated membranes extend over the viscera and muscle layer from a single giant *pseudocoelocyte* placed on the dorsal side where pharynx connects with the intestine. Four more star shaped giant cells lie in relation to epidermal chords. Thus in all there are five pseudocoelocytes. The pseudocoelocytes and fenestrated membranes may be regarded as a kind of mesenchyme.

In fact, the spaces are said to be large vacuoles in a few, enormous cells. A protein-rich fluid of characteristic, unpleasent odour fills the cavity. This fluid may produce some of the toxic and inflammatory symptoms of ascarid infections.

Locomotion

Locomotion is by means of longitudinal muscles which produce undulating movements of the body in the dorsoventral waves. The circular muscles are absent to counteract against longitudinal muscles, the hydrostatic pressure of body fluid against the cuticle serves the purpose.

DIGESTIVE SYSTEM

The alimentary canal of *Ascaris* is complete. It comprises mouth, pharynx, intestine, rectum and anus.

Mouth

It lies at the anterior end of the body. It is bounded by three lips-one mid-dorsal and two latero-ventral in position. It leads into the pharynx.

Pharynx

The pharynx is muscular, about 10 mm. long. It is provided with a triradiate, cuticulate lumen, and has a wall composed of radially arranged, lumen-dilating muscles. A single, pinnately branched dorsal, and a pair of palmately branched subventral, uninucleate glands are embedded in the wall of the pharynx. These are unicellular structures discharging their contents into pharyngeal cavity. The wall is composed of syncytial epithelium. The pharynx is followed by the intestine. The pharyngeal wall projects into the intestine as a pharyngeo-intestinal valve.

Intestine

It is a long, thin, walled, dorsoventrally flattened tube which is devoid of muscles. The intestinal wall is formed of a single layer of elongated columnar cells lined externally by a *basement membrane* and a thin layer

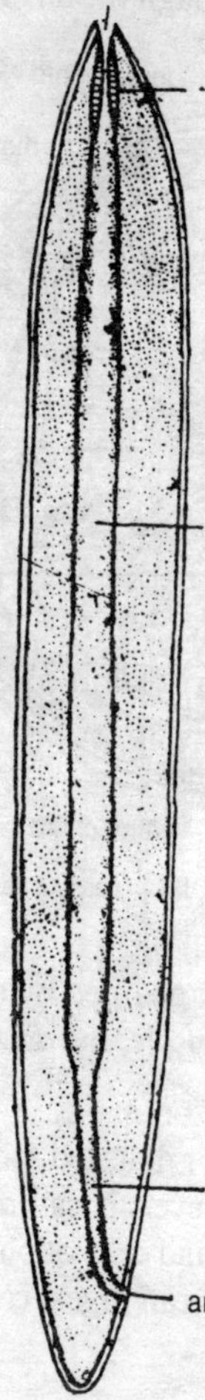

Fig. 9.7. *Ascaris.* Alimentary canal.

of *cuticle*. The free inner margins of columnar cells are produced into several finger-like projections called the *microvilli*. These form a *brush border* and increase the surface area for absorption.

Rectum

The intestine is followed by a short rectum which is also dorso-ventrally flattened. Wall of the *rectum* consists of tall columnar cells and is lined internally by cuticle and externally by muscle tissue. In male, the rectum opens into the *cloaca* which also receives the *ejaculatory duct*, while in female, it opens out through the *anus.*

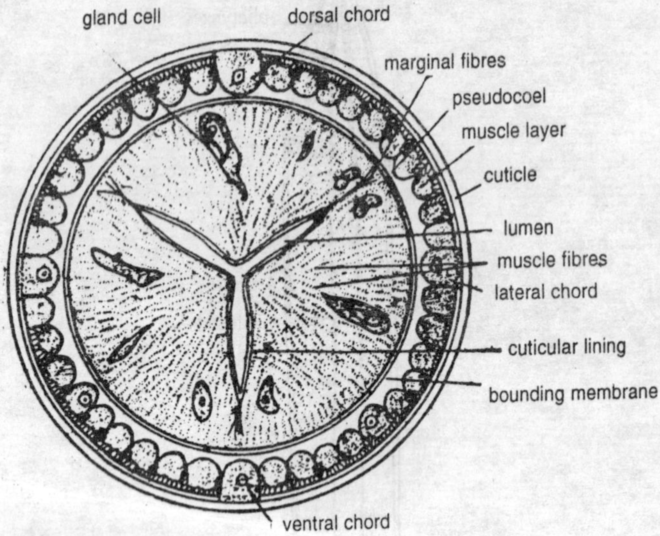

Fig. 9.8. *Ascaris.* T.S. body passing through phrynx.

Anus

The anus is a transverse slit guarded by anterior and posterior lips and is provided with a special muscle, the *depressor ani.*

Digestion

Ascaris gets semi-digested or digested food from its host. So it does not require to digest food. However, it can digest semi-digest food by means of enzymes secreted by gland cells of pharynx and cells of anterior part of intestine. These enzymes can digest carbohydrates, proteins and fats.

Absorption

The cells of intestine generally absorb digested food. Excess fat and glycogen are stored in different tissue.

Egestion

By the contraction of dilator muscles attached to rectum undigested food is expelled out.

RESPIRATION

There being no special organs of respiration, but the parasite carry on anaerobic respiration by breaking down glucose and forming carbon dioxide and fatty acid which are eliminated out through the cuticle. In the presence of oxygen, it respires aerobically. The small amount of haemoglobin in the body wall (muscles) and pseudocoel fluid takes up the oxygen even when it is present in low tension.

EXCRETORY SYSTEM

The excretory system of *Ascaris* is built entirely on a different plan. The flame cells or nephridia are absent. *Renette cells* of osmoregulatory function found in other nematodes are not found here.

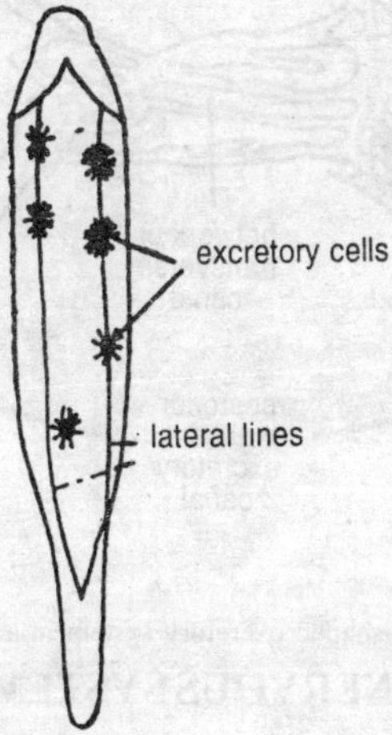

Fig. 9.9. *Ascaris.* Excretory system.

In *Ascaris*, the excretory system is made up of a syncytium containing four or more nuclei. It consists of:

1. Excretory duct

There is one excretory duct having two or more nuclei embedded into its walls. The excretory duct opens out through excretory pore.

2. Excretory cell

There is an excretory cell comprising two lateral excretory filaments running along each side of the body adjacent to the lateral chord.

3. Commissure

An excretory bridge is formed where two lateral filaments join. Each excretory filament contains an intracellular excretory canal. The intracellular excretory canals divide up into a network of canaliculi in the commissural region. Excretory products are believed to diffuse into the canals which secrete *urea*, which is eleminated out through the excretory pore.

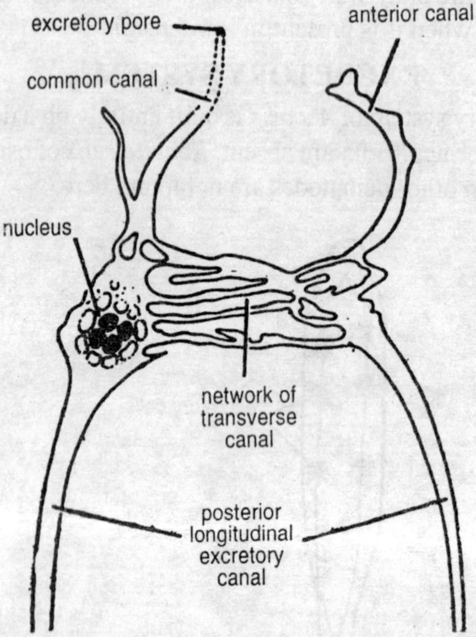

Fig. 9.10. *Ascaris.* H-shaped excretory system in anterior region of body.

NERVOUS SYSTEM

The nervous system consists of ganglia and nerves, all in close association with the ectoderm. The nervous system of *Ascaris* has been thoroughly studied by *Goldschmidt*. The number of cells taking part in the formation of entire system is apparently fixed and it is 162. There is a distinct *circumpharyngeal nerve ring* around the pharynx which connects a pair of *lateral ganglia*, a *dorsal ganglion*, a pair of *subdorsal ganglia* (each sub-divided into six smaller ganglia) and a pair of large *ventral ganglia.* From these ganglia of the nerve-ring there proceed backwards eight nerves: a *mid-dorsal nerve* that passes through the dorsal epidermal chord; a *mid-ventral nerve* that passes through the ventral epidermal chord; a pair of *dorsolateral nerves* that run just above the lateral chords; a pair of *lateral nerves* that extend through the lateral epidermis chords and a pair of *ventro-lateral nerves* that run just below the lateral epidermal chords. Of these, the ventral nerve is most prominant and is ganglionated through out its length, therefore, it is regarded as nerve cord. Posteriorly the nerve cord joins a *pre-anal ganglion.* The dorsal and ventral nerves are connected at intervals by asymmetrical *dorso-ventral connectives* throughout their length. Similarly, the ventral and lateral nerves are also connected at interval by *ventro-lateral commissures.*

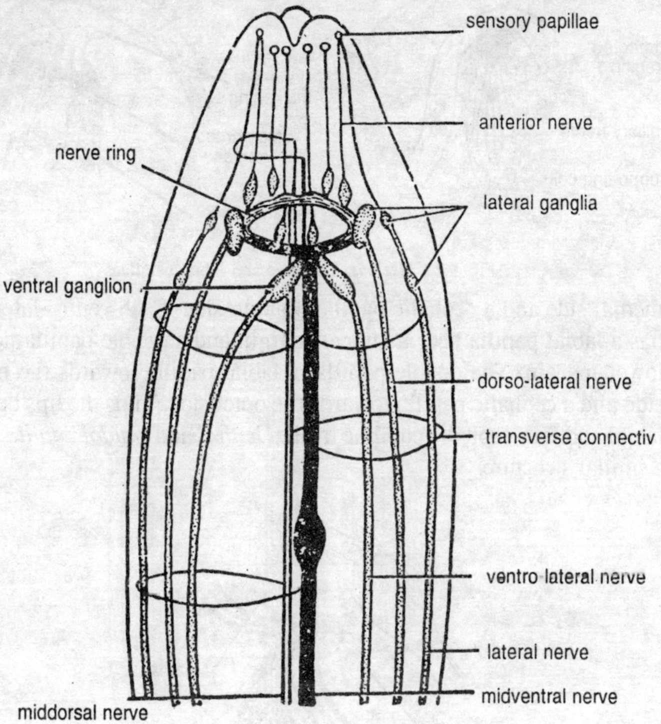

Fig. 9.11. *Ascaris:* Nervous system.

One of the six ganglia formed by sub-division of each lateral ganglion is known as *amphidial ganglion* which supply nerves to amphid of its side. The genital papillae of the male worm are innervated from special *bursal nerves.*

SENSE ORGANS

The sense organs include papillae, amphids and phasmids.

Papillae

Various types of papillae such as *labial, cervical* and *genital papillae* are present in *Ascaris.* A labial papilla consists of a nerve axis placed inside a cuticular elevation surrounded by supporting cells. The dorsal lip has two double papillae, each consisting of a labial papilla on the

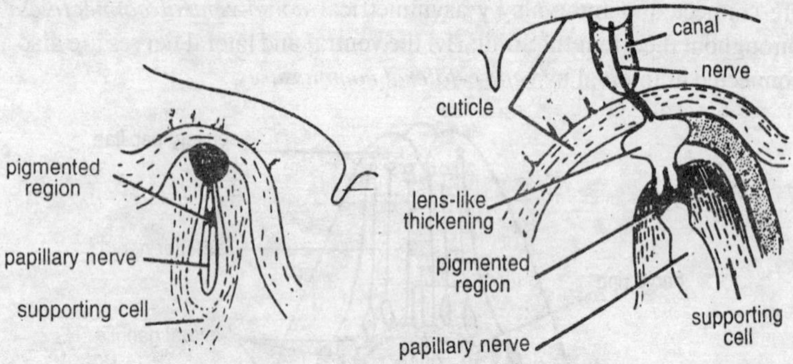

Fig. 9.12. *Ascaris.* Cephalic and labial papillae.

medial side and a cephalic papilla on outerside. Each ventro-lateral lip has a labial papilla near its upper margin and a double papilla near its lower margin. The double papilla, a labial papilla towards the medial side and a cephalic papilla towards the outer side. Thus the lips bear six labial and four cephalic papillae in all. *Genital* and *caudal papillae* have a similar structure.

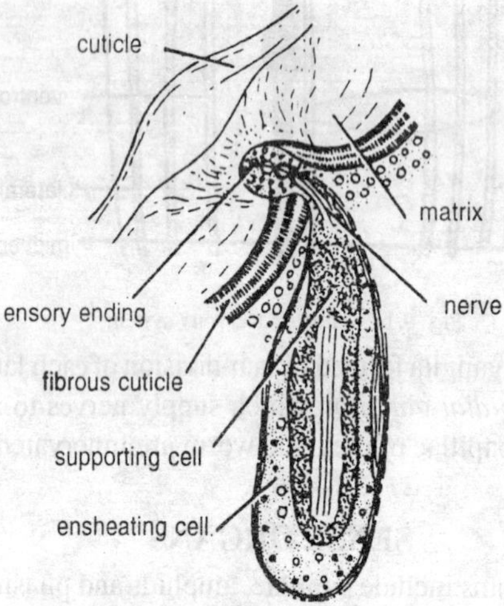

Fig. 9.13. *Ascaris.* Cervical papilla.

Amphids

These are two in number, one on each *ventro- lateral lip,* a little above the labial papilla. An amphid consists of supporting cells opening out side by a pore and having a bundle of nerve-fibres that unite into one fibre. These are probably a chaemoreceptor.

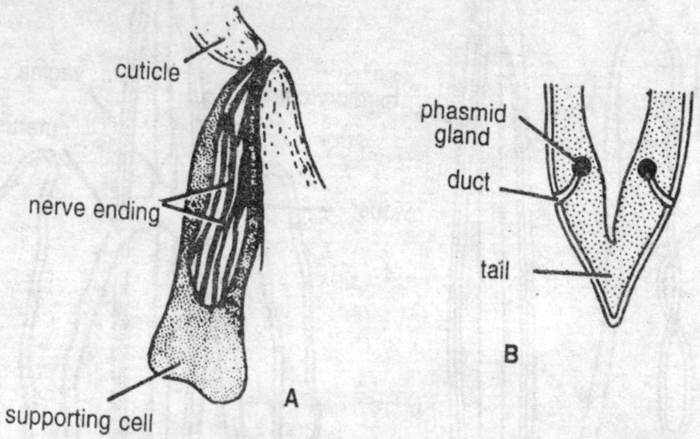

Fig. 9.14. *Ascaris.* A—Amphid, B—Phasmid.

Phasmids

A pair of *phasmids* occur on each side of the tail. Phasmid is the outlet of a unicellular gland opening over a small papilla which is probably chaemoreceptor.

REPRODUCTIVE SYSTEM

Sexes are separate and there is a well distinct sexual dimorphism.

1. Male Reproductive System

The male reproductive organs are :

1. ***Testis.*** The testis is a long thin, coiled and thread-like structure in the anterior part of pseudocoel. It represents *monorchic condition* (presence of single testis) and *telogonic* mode of origin of germ cells (i.e. germ cells developing only from the proximal part of gonad). The testis can be distinguished into three parts:-

   (a) anterior solid part consisting of sex cell, (b) middle part and (c) posterior part. The central axis of the tube is in the form of a solid *cytoplasmic rachis*, around which the clusters of amoeboid sperms in various stages of development are present.

   Testis is continued posteriorly into the vas deferens.

2. ***Vas deferens.*** The vas deferens is as wide as the testis and the two cannot be marked externally. It is very short and leads into the seminal vesicle or Vesicula seminalis.
3. ***Vesicula seminalis.*** The vesicula seminalis is a wide tube present in the posterior 1/3rd part of the body. It joins the ejaculatory duct.

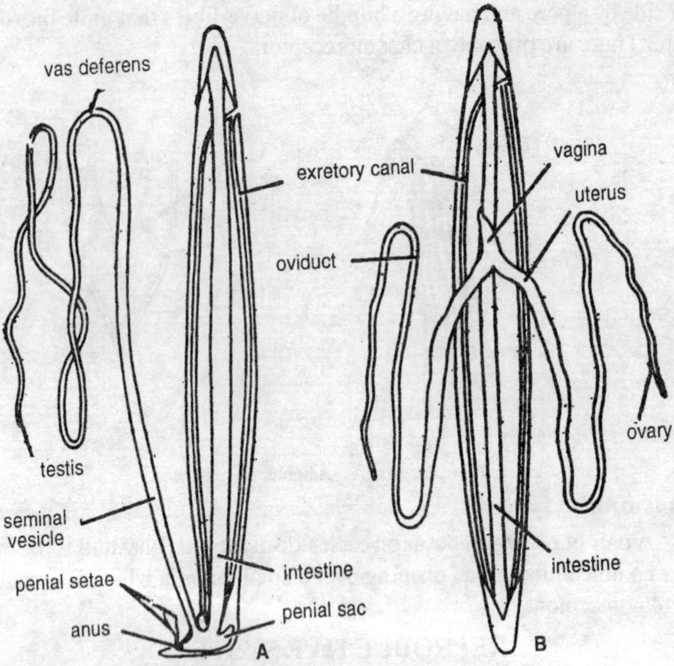

Fig. 9.15. *Ascaris*. A—Male reproductive system, B—Female reproductive system.

4. ***Ejaculatory duct***. It is a short, narrow but muscular tube opening into the cloaca, which leads outside by cloacal opening.
5. ***Penial sacs with penial setae.*** On the dorsal side of the cloaca there is a pair of sac-like structures called *penial sacs.* These are formed by evaginations of the cloaca. Each pouch secretes and lodges a club-shaped *spicule* or *penial seta*. The spicule is enclosed in a sheath and consists of a cytoplasmic core surrounded by thick cuticle. Each spicule measures 2-3.5 mm. in length. The penial sac have special *protractor* and *retractor muscles* from their walls,the muscles push the spicules out and withdraw them back into the pocket.

II. Female Reproductive Organs

The female reproductive organs include:

1. ***Ovaries.*** The ovaries represent *didelphic* condition i.e. there are two ovaries. Each ovary is a long, thread-like, much twisted and blind tubule occupying posterior 1/3rd of the body. Its wall consists of a single layer of cuboidal epithelial cells lined externally by basement

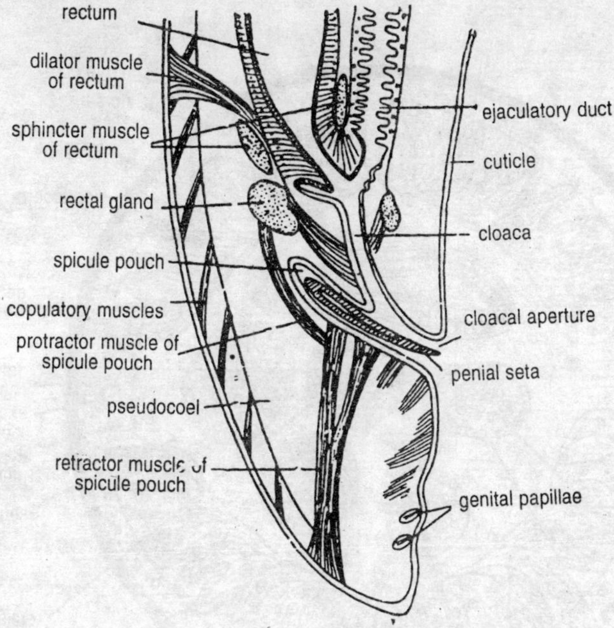

Fig. 9.16. *Ascaris*. L.S. through posterior end of male.

membrane. There is no lumen, the central part is occupied by the *cytoplasmic rachis,* around which young eggs are clustered.

2. ***Oviducts.*** The oviducts are two thin ducts very similar to the ovaries. These open into the uteri.
3. ***Uteri.*** Each oviduct continues into a much wider, thicker and almost untwisted tube, the uterus. The wall of the uterus is composed of tufted cells, surrounded by a muscle layer with inner *circular* and outer *oblique fibres*.
4. ***Vagina.*** Two uteri join to form one vagina which opens to the outside by female genital pore or *vulva,* lying on mid-ventral side about 1/3 of the length of the body posterior to anterior end.

LIFE-CYCLE

Ascaris is *monogenetic* i.e., it requires only one host to complete its life-cycle.

1. Copulation and Fertilization

Ascaris copulate in the intestine of man. During the process sperms are ejected by male into vagina of the female. These move in the uterus

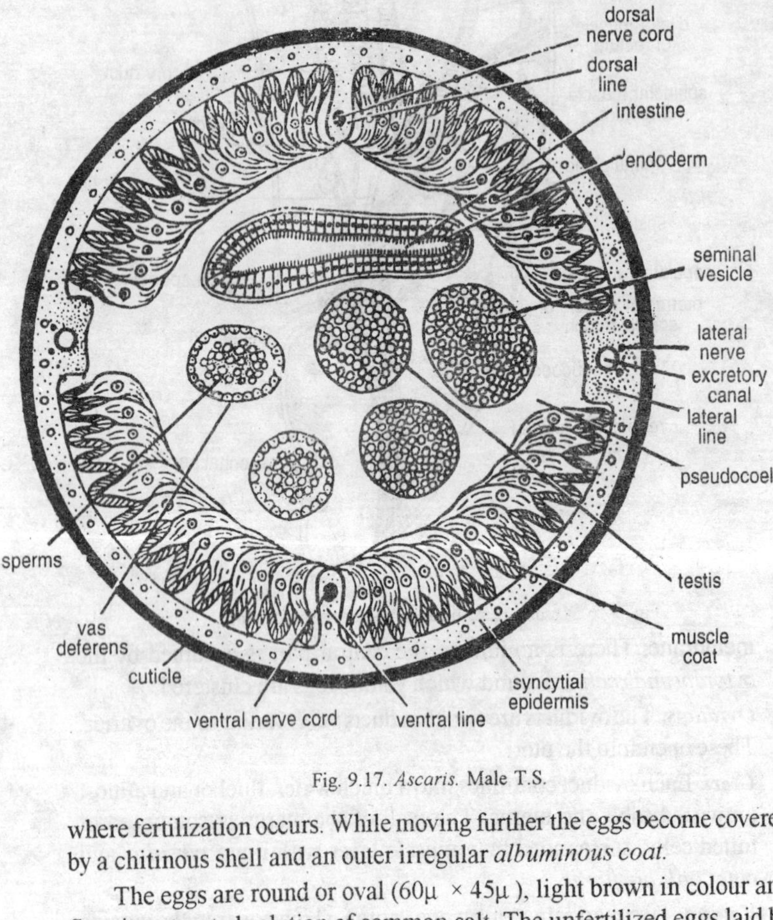

Fig. 9.17. *Ascaris.* Male T.S.

where fertilization occurs. While moving further the eggs become covered by a chitinous shell and an outer irregular *albuminous coat.*

The eggs are round or oval (60μ × 45μ), light brown in colour and float in saturated solution of common salt. The unfertilized eggs laid by the mother, may mix up with the fertilized eggs but the unfertilized eggs are larger (90μ × 45μ) and sink in saturated salt solution.

2. Egg-laying

The uteri of a single female may contain as many as 2,70,000,00 eggs (*Cram*, 1927), with an average daily production of about 2,00,000 eggs. The fertilized eggs are released in the intestine of host through the vulva. From the intestine of host they are eliminated along with faecal matters.

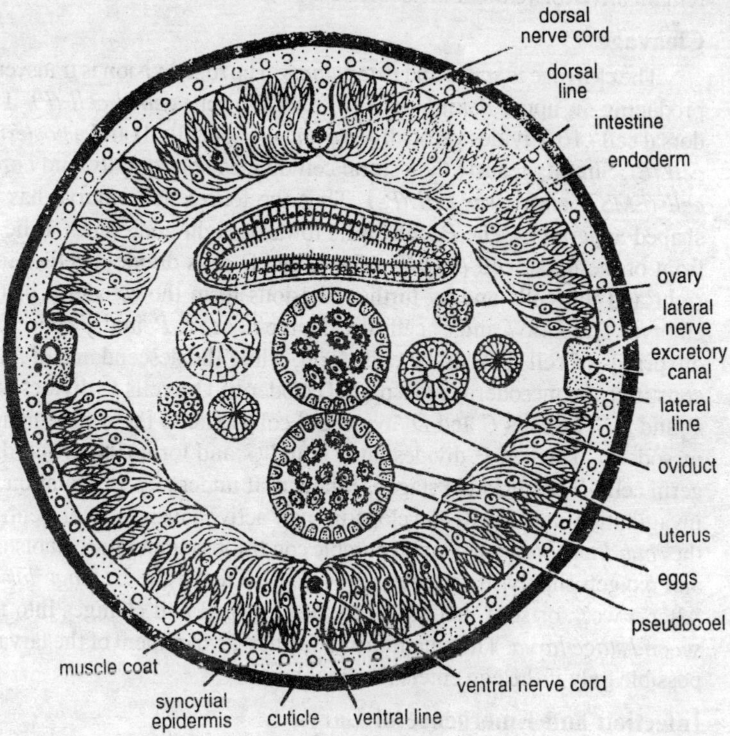

Fig. 9.18. *Ascaris.* Female T.S.

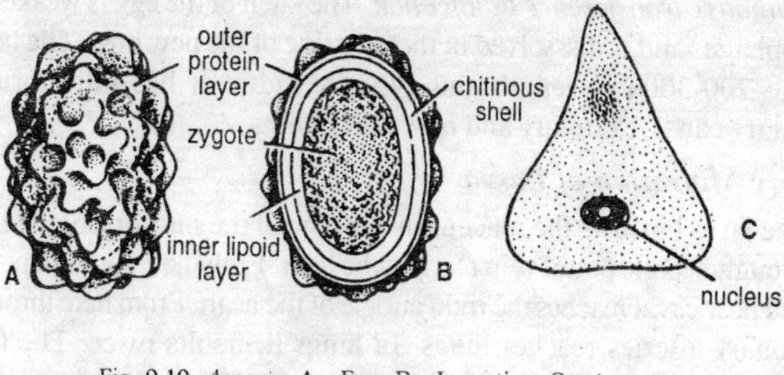

Fig. 9.19. *Ascaris.* A—Egg, B—In section, C—A sperm.

3. Development

The development takes place at 85°F i.e. outside the body of man. It is stopped at 60° F and eggs gradually degenerated at 100° F. They can remain alive for years in moist soil.

Cleavage

The cleavage is *spiral* and *determinate*. The first division is transverse producing an upper *dorsal cell (AB)* and a lower *ventral cell (P)*. The dorsal cell *(AB)* divides vertically into an anterior cell *(A)* and a *posterior cell (B)*. Simultaneously the ventral cell divides horizontally into *upper cell (EMST)* and *lower cell* (P_2). Thus the four-celled embryo has T-shaped apperance. The P_2 cell shifts towards right and comes to lie in front of the upper cell *(EMST)*. The A and B cells divide together into right and left cells and by further divisions form the ectoderm of the embryo. The *EMST* and P_2 cells divide into *E, MST,* P_3 and *C* cells. The offsprings of cell *E* form the endoderm, while the descendants of *MST* constitute the mesoderm and a part of ectoderm. The cells P_3 divides into P_4 and *D*. The cells *C* and *D* divide and contribute to the ectoderm and mesoderm. The cell P_4 divides into G_1 and G_2 and forms the primordial germ cells. Thus blastula stage is attained. It undergoes gastrulation by invagination and finally develops into an active juvenile, representing the *rhabditoid stage*. Under favourable conditions temperature, moisture and oxygen supply, it requires 10-14 days to reach the *Ist stage larva*. After a week or so it moults inside the egg shell and changes into the *second stage larva*. The hatching and further development of the larva is possible only if the egg enters the final host.

Infection and Emergence of Larva

The eggs containing *rhabditiform larvae* contaminate food and water. If someone takes this contaminated food or drinks contaminated water the eggs reach the healthy man. This type of *transference* is called *contaminative transference* or *infection*. The shell of the egg is weakened in oesophagus and is disssolved in the intestine of the new host. The larva is about 200-300μ in length and 14μ in width. It has fully formed alimentary canal, excretory and nervous systems.

Primary Migration of Larva

The larva burrows the mucous membrane of the intestine of the host and through hepatic portal vein reaches the liver. From here through portal vein and post-caval reaches the right auricle of the heart. From here through pulomonary arteries reaches lungs. In lungs it moults twice. The first moulting takes place after 5-6 days (*third stage larva*) and second moulting after 10 days forming *fourth stage larva*. This larva now increases in size and is about 2 mm. in length.

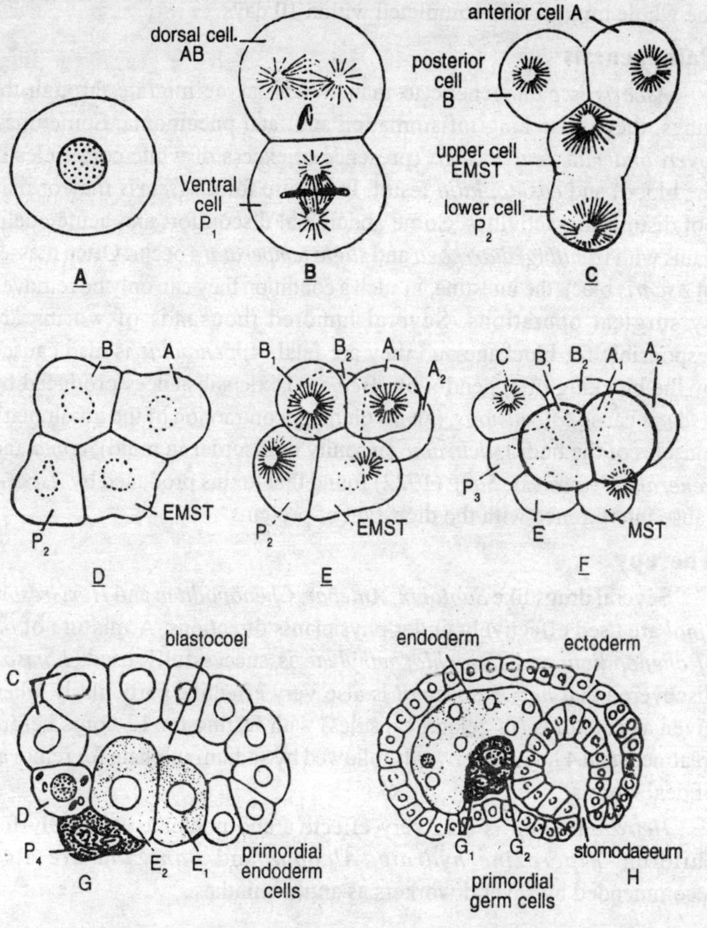

Fig. 9.20. *Ascaris.* Stages showing early development. A—Zygote, B—2-cell stage, C—4-cell stage (T-shaped), D—4-cell stage (rhomboidal), E- 6-cell stage, F—8-cell stage, G—Median sagittal section through blastula, H—Median section through the embryo after invagination of stomodaeum and the primordial germ cells.

Secondary Migration

From lung the larva reaches bronchii, trachea, larynx and finally to pharynx where it irritates causing coughing by which it re-enters the alimentary canal when the cough is swallowed. After reaching the intestine through stomach it moults fourth or last time. The young *Ascaris* becomes adult sexually mature within 6-10 weeks and begins its life-cycle again, the whole migration is completed within 10 days.

Pathogenesis

Ascaris is pathogenetic to man. When larvae migrate through the lungs, they cause lung-inflammation and fatal pneumonia. Sometimes, *fever, anaemia, leucocytosis* (presence of excess of white corpuscles in the blood) and *eosinophilia* result. In mature form, *Ascaris* may or may not disturb life-activities. Some abdominal discomfort and acute coelic pains with *vomiting, diarrhoea* and *slight temperature* occur. Often masses of *Ascaris* block the intestine, in such a condition they can only be removed by surgical operations. Several hundred thousands of worms are responsible for blockage and they are fatal. *Appendicitis* is also caused by the blocking of appendix by *Ascaris*. Toxic substances produced by *Ascaris* cause *convulsions*, (an involuntary contraction of the involuntary muscles of the body), *delirium*, (insanity or disorder in mind), *coma* and *general nervousness*. *Sang* (1922) found that toxins produced by *Ascaris* cause interference with the digestion of proteins.

Therapy

Several drugs like *Santonin, Antepar, Chenopodium* and *Hexylresorcinol* are used effectively under physician's directions. A mixture of *oil of chenopdium* and *tetrachloroethylene* is successfully used. *Lamson* discovered that *hexylresorcinol* is also very effective particularly when given as crystoids (in gelatin capsules) with fasting for 12 hours before treatment and 4 hours afterwards followed by sodium sulphate for removal of deal worms.

Hetrazan syrup is also very effective and nontoxic especially for children. *Piperazine hydrate, Alcopar* and *Banocida* are also recommended by several workers as antihelminthes.

Prevention

Adequate sanitation plus education should prevent *ascariasis*. Lack of adequate fertilizer in many parts of the undeveloped world means that night soil must be used to fertilize crops. Due to the resistance of *Ascaris* eggs to standard disinfectants, the situation would appears to be hopeless. There is a realistic solution. *Thitasut* (1961) reported that aqueous solution containing 120 ppm free iodine killed *Ascaris lumbricoides* eggs within 5 min or less and did not effect the flavour of leafy vegetables.

One of the problems in control is preventing children from ingesting eggs with contaminated soil. Given the habits of children throughout the world, including dirt eating, playing in contaminated soil and then placing their fingers in their mouths, *ascariasis* will unfortunately be with us until adequate sanitation is accomplished.

Revision Questions

1. Describe the morphology, histology and physiology of alimentary canal of a round worm studied by you.
2. Describe external features, life-history and economic importance of *Ascaris*.
3. Give an account of structure and life-history of *Ascaris*.
4. Name any three nematodes, which are harmful to man. Describe the life-history of any one of them.
5. Draw fully labelled diagrams. Description is not required : (a) T.S. through male *Ascaris*. (b) T.S. of female *Ascaris*.
6. Give in brief the structure and life-history of *Ascaris*. Suggest suitable control measures.
7. Give the salient features of a parasitic round worm found in human intestine.
8. Describe the reproductive organs of *Ascaris* and explain how infection in a new host takes place.
9. Give an account of morphology and physiology of *Ascaris*.
10. Write short notes on : (a) Pseudocoel, (b) Muscle cells, (c) Aphids, (d) Phasmids.

10

Ancylostoma

No group of the nematodes causes more injury to man or greater economic loss through attacks on his domestic animals then the members of the order Strongyloidea, the great majority of which are parasites of mammals. Many are blood suckers and cause severe injury to their hosts by loss of blood sucked by them or wasted from hemorrhages, the result is anemia.

SYSTEMATIC POSITION

Phylum	-	Aschelminthes
Class	-	Nematoda
Order	-	Strongyloidea
Genus	-	*Ancylostoma*

Distribution

Human hookworm is endemic in warm, wet climates, but each species has some limitation on its distribution.

Ancylostoma duodenale is the common hookwork of man in tropical, subtropical and temperate regions of Asia, Africa, Pacific islands and South America. The other important human hookwork is *Necator americanus,* more common in the United States. Some other species of *Ancylostoma,* namely, *A. caninum* and *A. braziliense*, both in dogs and cats, and *A. malayanum* in bear, are also common.

Habits and habitat

A. duodenale is the principle hookworm that infects humans. Following early morning, they are found in the small intestine lumen. Hookworms sometimes mature in abnormal hosts, and hosts other than man can be infected with human hookworms.

MORPHOLOGY

The name 'hookworm', is rather misleading because there is absence of hooks. The anterior hook-like bend probably is responsible for the common name, hookworm.

Shape and Size

The sexes are separate. It is a cylindrical worm and whitish or grey in colour. The females are longer than males measuring 10-13 mm. in length and 0.6 mm. in width. The males measures 8-11 mm. in length and 0.45 mm. in width.

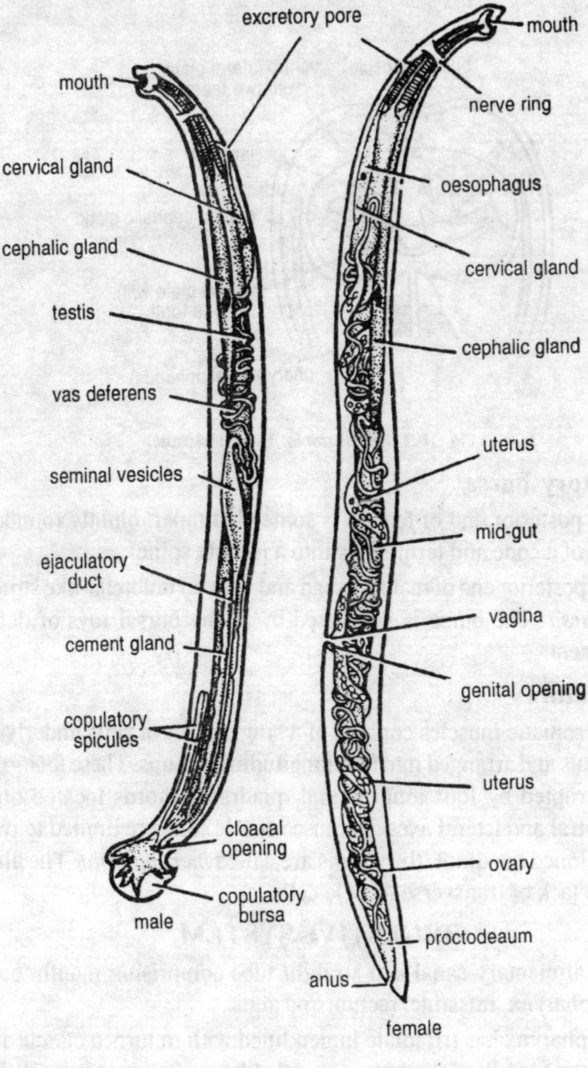

Fig. 10.1. *Ancytostoma duodenale.*

Buccal capsule

The anterior end of male and female is similar, somewhat narrower and curved dorsally. It bears a large cup-shaped buccal capsule for attachment to the intestinal wall. The buccal capsule has two *chitinous plates,* bearing teeth. The *teeth* make a tiny cut in the intestinal wall of the host and hoolk-worm sucks blood by the sucking action of pharynx.

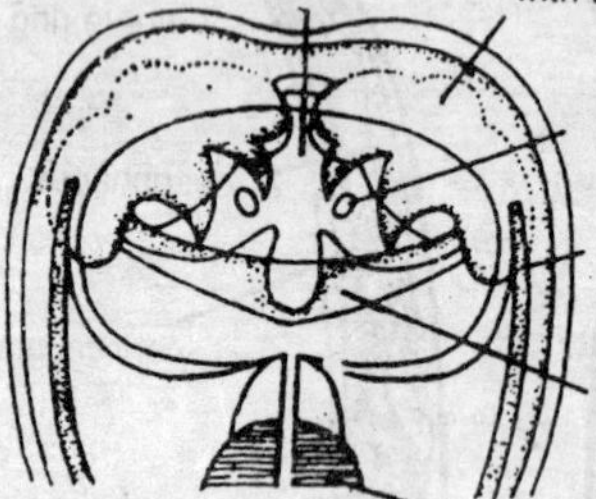

Fig. 10.2. *Ancytostoma.* Buccal capsule.

Copulatory bursa

The posterior end of female is somewhat taper, bluntly rounded in the form of a cone and terminates into a minute spine.

The posterior end of male is broad and with an umbrella-like structure called *bursa.* The bursa is supported by fleshy bursal rays of definite arrangement.

Musculature

The somatic muscles consists of a single layer of cells underlying a hypodermis and arranged into four longitudinal groups. These four groups are interrupted by four longitudinal quadratal chords located on the dorsoventral and leteral axes. Because muscle cells are limited to two in each sectioned quadrant, the worms are called *meromyarial.* The almost complete lack of transverse muscle cells.

DIGESTIVE SYSTEM

The alimentary canal is a straight tube comprising mouth, buccal capsule, pharynx, intestine, rectum and anus.

The pharynx has triradiate lumen lined with in turned cuticle and a wall composed of strong radiating muscle fibres extends posteriorly from the buccal capsule about 1/6 the body length. The posterior region of pharynx is slightly dilated and embedded in its wall are three pharyngeal glands. One is median dorsal in postion and a pair located subventrally. These glands open into the pharynx near the nerve ring. An intestine begins immediately behind the triradiate pharynglal valve and proceeds as a straight tube to a short rectum. It is composed of a single layer of columnar cells. In the intestine the blood of host and other tissues are digested.

EXCRETORY SYSTEM

The excretory system is rather complicated. Flame cells are absent in nematodes. The excretory system, consists of two excretory canals embedded in the lateral chords and an isthmus in the pharyngeal region. Surrounding or adjacent to the isthmus, are other excretory structures including an excretory sinus into which tributaries from the lateral canals open and from which ventrally is given off a short, cuticulate, terminal duct. The duct opens at a ventral excretory pore. From the excretory sinus two large, *subventral glands* of uncertain function arise posteriorly. The tributary canals emerge from subventral glands and penetrate the lateral chords where they join the lateral canals. These lateral canals extend from the region of buccal capsule to the posterior end of the body where they end blindly.

NERVOUS SYSTEM

Encircling the pharynx, in front of the excretory pore, is a *nerve ring*. From a large ganglion located on the ventral portion of the nerve ring, a large nerve runs in the ventral chords to the posterior end, and smaller nerves run in the lateral chords anteriorly and posteriorly. Four small nerves run forward from the ring to supply anterior end. These main nerve trunks are connected in a complicated fashion by nerve commissures.

REPRODUCTIVE SYSTEM

The male and female reproductive system are as follows :

Male Reproductive Organs

The tubular *testis* occupies the middle third of the body. It begins at the junction of the middle and posterior body third and runs in a random fashion to the middle of the body, from which point forward it coils loosely about the intestine. It leads posteriorly into *vas deferens* followed by *seminal vesicle*. Seminal vesicle which connects posteriorly to an ejaculatory duct leading directly to the cloaca. The ejaculatory duct, through most of its course, lies between two large, multicellular cement glands whose secretion apparently serves as an adhesive material holding the pair of worms together at copulation.

The posterior end of male bears a wide *bursa*, two *copulatory spicules* and a *gubernaculum*. The bursa is a long, wide, bilaterally symmetrical structure, with supporting, paired, fleshy rays having a specific distribution. The two spicules are long and slender. Each measures 2000 μm. They are protruded and retracted by special muscles and help in the transport of sperms to the vulva of female. The gubernaculum is a hard piece present on the dorsal wall of cloaca. It is said to be used to guide the spicules.

Female Reproductive Organs

Two much convoluted *ovarian tubules*, one anterior and other posterior to the level of gonopore are present. From the ends of the ovaries, nearest the vulva, short, narrow *oviducts* carry the ova into wider *seminal receptacles* where fertilization occur. From each seminal receptacle arises a muscular *uterus* that terminate into *vagina*. Two vaginae join and then open to the surface at the vulva.

LIFE-CYCLE

Copulation and Fertilization

Fertilization is internal and during copulation the bursa of male is applied on the vulva of female. The female *Ancylostom*a lays about 20,000 eggs per day. The egg is about 60 x 30μ . These are passed along the host faeces at four-celled stage and do not develop further until exposed to air. Under favourable conditions the embryo hatches out of the egg in less than 24 hours in the faeces contained in the mud.

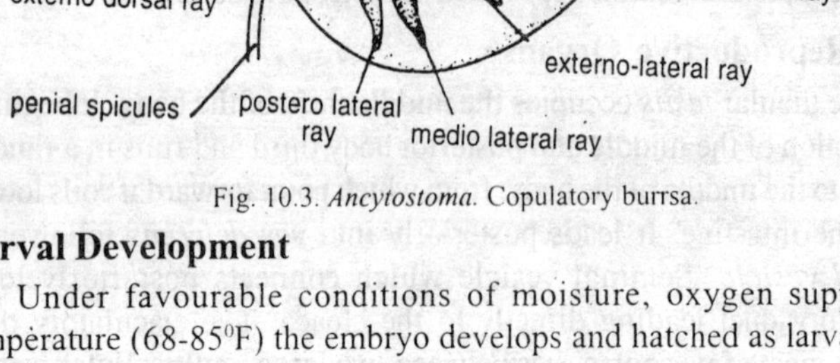

Fig. 10.3. *Ancytostoma.* Copulatory burrsa.

Larval Development

Under favourable conditions of moisture, oxygen supply and temperature (68-85^{0}F) the embryo develops and hatched as larva.

The hatched larva is the first larval stage or the first *'rhabditiform larva'*. It has mouth, pharynx and a simple intestine. It feeds on bacteria and organic debris in the faeces and grows rapidly. At the end of 2nd day it undergoes first moult and changes into second stage larva (or the *strongyliform larva*). At about 5th day it undergoes second moult to become *filariform juvenile*. This is the infective non-feeding stage. The skin of the second stage larva is retained as a protective sheat. The larvae live in the upper half inch of soil.

Infection

The third stage or the filariform juvenile represents the infective stage. It may remain alive in the moist soil for nearly four months without feeding. When a man walks bare-footed on the infested soil, the larva penetrates the skin. Infection can also occur by ingestion of infective juvenile in contaminated food or water.

Larval Migration

Within 24 hours of infection, the larvae reach the blood-vessels. From here they are carried first to the right ventricle of the heart and then to the lungs. From lungs the larvae assend the trachea and to the pharynx. In pharynx they are swallowed through oesophagus and finally reach to the small intestine. In the ileum they become attached to the mucous lining and feed on blood of the host. In 5-6 weeks, they moult twice to become adult and begin to reproduce. The normal life-span of hookworm is 5-7 years.

Hookworm-Pathology

Hookworm infection gives rise to a wide range of signs and symptoms indicative of a number of pathological conditions loosely termed 'hookworm disease'. The hallmark of hook-worm disease is iron-deficiency anaemia, a condition which should give no cause for concern when nutrition is adequate, or could be readily reversed by iron therapy. Sadly, in hookworm-endemic areas adequate nutrition is the exception, and not only is dietary protein in short supply but it is usually of poor biological value. similarly, dietary iron supplementation is not readily available, making it difficult for the subjects to maintain the required iron balance. It is for these reasons that iron-deficiency anaemia due to hookwork assumes the importance it does in the tropics. *Miller* (1979) has given a comprehensive review of hookworm infection. A drastic summary is presented here. For greater details, the reader should consult Miller's review.

Morbidity in hookworm infections can be either acute or chronic : acute infection is due to the activities of migrating larvae and the adults recently reaching the intestine, while morbidity due to chronic infections arise from the activities of established adults in the intestine as well as the consequences of the physiological, biochemical, and haematological disturbances they cause.

Signs and symptoms of acute infection

Penetrating larvae produce inflammation (dermatitis) which is intensified by secondary bacterial infection causing an irritation of varying severity known as *ground itch.* They do not, however, produce the serpentine tracks associated with cutaneous larva migrans characteristic of other non-human hookworm larvae. Although there are no marked pulmonary signs, there are clear signs of supper respiratory tract involvement expressed as laryngitis, pain when speaking and swallowing etc.

On arriving in the intestine the worms induce signs and symptoms including severe abdominal (epigastric) pain, loss of appetite, nausea and vomiting, flatulence, etc. Associated with abdominal pain are indigestion and diarrhoea produced on arrival of the fourth stage larva, and dysentery and constipation arising from the blood-sucking activities of adult worms in the intestine. But the most distinct sign at this stage is anaemia which, when there are adequate iron reserves, need constitute no serious problem. The acute haemorrhage which frequently results in fatal anaemia in canine hookworm infection has only rarely been reported in humans. This condition, 'acute uncompensated haemorrhagic anaemia', to give it its full name, may be more common than is generally thought, particularly in remote and backward areas where high infant mortality is not routinely reported and fully documented.

Signs and symptoms of chronic infection

Signs and symptoms of chronic infection can be classified broadly into two :

1. *Primary*, which are the direct results of the actions of the worms themselves. These signs recede or are greatly reduced in intensity when the worms are expelled.
2. *Secondary*, which arise indirectly from the effects of the anaemia caused by the worms. It is possible to remove these signs without expelling the worms, by restoring the haematological level with iron therapy.

Primary signs of chronic infection

These are essentially the same as in acute infection—abdominal pain, diarrhoea, and indigestion, constipation, and a desire to eat soil (geophagy). However, in endemic areas where there is continuous reinfection, these signs may be due to new infection.

Secondary morbidity in chronic infection

The main condition here is iron-deficiency anaemia which develops slowly and insiduously and may not become apparent for years, especially with light infections and adequate iron intake or iron reserves. Of particular significance, the condition can be effectively managed, corrected, or prevented altogether by suitable dietary intervention, especially by the administration of supplementary iron. *Foy* and *Nelson* (1963), *Gilles, Waston Williams*, and *Ball* (1964), and *Roche* and *Layrisse* (1966), have given good accounts or iron-deficiency anaemia. Unfortunately, the following complicating intercurrent conditions have been known to hamper the effectiveness of iron therapy :

1. Dietary deficiency
2. Interference with iron absorption
3. Kwashiorkor
4. Chronic bacterial infections like tuberculosis and syphilis.
5. Parasitic infections like amoebiasis, malaria, and schistosomiasis.

In addition to iron-deficiency anaemia in chronic hookworm infection, the following conditions also develop : hypoproteinaemia, hypoalbuminaemia, and oedema. Other secondary signs and symptoms include cardiac insufficiency and congestive cardiac failure, weakness, breathlessness, palpitation, systolic murmur, angina pectoris, vertigo, fainting, etc.

However, just as serious as the effects of anaemia and other conditions are the deficiencies apparent in the health profiles of individuals and communities ostensibly induced by hookworm disease. Thus :

1. Productivity is greatly undermined because of disinclination and inability to work.
2. Hookworm subjects tend to present an image of dullness, stupidity, confusion, poor memory, etc.
3. The effects on children are more severe. Chronic infection has been held responsible for the stunting of the physical and mental development of children; indeed, such children have been reported to be the most stupid in their class in endemic areas.

4. Hookworm disease has been incriminated in many reproductive disorders including;

(a) Delayed puberty.

(b) Miscarriages, abortion, stillbirths.

(c) Up to 70 per cent of foetal, neonatal, and early infant mortality, as well as 25 per cent of maternal mortality in endemic areas have been attributed to hookworm disease.

Anaemia due to hookworm is one of the commonest causes of hospitalization in tropical Africa and is one of the principal causes of heart disease. However, the severity of the anaemia depends on the worm load and the dietary intake of the patient. For example, in Nigeria, it was estimated that 800 worms (with egg counts of 20,000 ova per gram of faeces) produced significant clinical anaemia, while in East Africa and Gambia, smaller worm burdens were responsible. This disparity is apparently due to the higher iron content of the Nigerian diet estimated at 40 mg per day. Of this amount, however, only about 5 per cent of the iron is absorbed and since normal daily requirement of iron is in the range of 0.8 to 3.2 mg, any reduction of the absorption below 5 per cent could upset the iron balance adversely. Also, the degree of morbidity depends on the species of hookworm and this is probably related to the amount of blood loss caused. Thus *A. duodenale* infections (with 0.15 ml per worm per day of blood) are more severe than *Necator americanus* infections responsible for only 0.03 ml per worm per day of blood loss.

One study in Northern Nigeria must be pointed out. Among the Fulanis where iron-deficiency anaemia is rampant, *Oomen* (1975) showed that the mild degree of hookworm infections meant that hookworm cannot be held responsible as a major contributory factor.

Mortality

Mortality due to hookworm disease has not been widely reported. However, *Bwimbo* (1970) reported that hookworm anaemia is the third most frequent cause of death in 1 to 5 year old children in Uganda, while *Trowell* (1956) and *Fowler*, *Knight*, and *Patel* (1968) consider that death from cardiovascular inadequacy and circulatory collapse brought about by hookworm disease is not uncommon. Finally, *Miller* (1979) is convinced that the incidence of hookworm mortality may be higher than generally accepted to be the case. This is particularly so in cases of infant and child mortality in many endemic areas which are neither reported, investigated, nor documented.

Pathophysiology

A comprehensive review has been given by *Roche* and *Layrisse* (1966). Pathophysiology is presented in the form of blood loss (enterorragia) accompanied by iron loss. Blood is lost;

1. When worms suck blood required for their own nutrition and respiration. It appears that they suck more blood than is actually necessary to satisfy their needs.
2. When blood is spilled around the worm's head.
3. When blood leaks from damaged blood vessels in the intestinal lining.

The amount of blood loss is estimated by using the technique of radioactive labelling of circulating erythrocytes with cromium. With this technique, it has been estimated that up to 200 ml per day may be lost by patients with heavy infection. The amount of iron lost is also important. This, however, depends on the amount of blood lost as well as the haemoglobin values at the time bleeding occurred. About 40 per cent of the iron in the blood lost may be reabsorbed although confirmatory proof is required for most reports.

Another pathological feature is haemolysis (erythrocyte destruction) caused by haemolysins produced by the hookworms.

On the whole, therefore, iron-deficiency (hypochromic) anaemia is the most important factor in the morbidity and mortality of chronic hookworm disease. This condition can be reversed by iron therapy and, in any case, should have little chance of developing in the first place if there is an adequate iron-rich diet or if supplementary dietary iron is introduced into the regular diet to restore the iron balance. As it happens, hookworm endemic areas tend to coincide with areas where iron deficiency is rampart, so the people ordinarily have a hard time achieving the minimum iron and blood levels, let alone having to content with heavy hookworm burdens. *Miller* (1979) has given an excellent summary of iron balance in relation to hookworm infection.

Hookworm—Diagnosis and Treatment

The specific diagnosis is finding the characteristic eggs in stool samples. It is often necessary to estimate the output of eggs as a mesure of the worm burden, and techniques for this are described by *Mardsen* (1977). It is usually not possible to distinguish between eggs of *N. americanus* and *A. duodenale* and so they are generally reported as hookworm eggs.

A sensitive method for hookworm identification is the 'test-tube filter-paper culture method'. Here, faeces is smeared at one end of a strip of filter paper, while the other end is dipped into water in a test-tube. The larvae which subsequently accumulate in the water can be identified and estimated. This method not only makes it possible to separate larvae of hookworm and *Strongyloides*, but to distinguish between the two species of hookworm as well.

Intradermal tests have been developed using various hookworm antigens to produce hypersensitive reactions. This is useful for mass screening for hookworms in endemic areas. The general appearance of a person suffering from hypochromic anaemia in a tropical area should serve as a clue.

Light infections require no treatment beyond iron therapy and indeed severe symptoms may not even develop if an iron rich diet is available. In heavy infections, however, supplementary dietary iron will alleviate the symptoms—if not reverse them altogether. chemotherapeutic agents include tetrachlororethylene, which is an effective and cheap hookworm expeller, particularly good for mass treatment. However, it causes *Ascaris lumbricoides* to knot together causing intestinal blockage with serious consequences. Bephenium hydroxynaphthoate (Alcopar) and mebendazole are preferable in cases of concurrent infections of hookworms and roundworms, being effective against the two.

There is no conclusive evidence that man can develop a functional protective immunity to hookworm infection. *Miller* (1979) is convinced that more research in this field may reveal immuity similar to that operating in the dog-hook-worm relationship.

Hookworm—Epidemiology

The epidemiology is influenced by the standard of sanitation and the existence of conducive environmental conditions for the development of the eggs and survival of the larvae. The features whch promote transmission, as well as those which are unfavourable, are well recognized. The following is a summary in outline of the principal factors gleaned mainly from the succinct review of *Schmidt* and *Roberts* (1981).

Chief among the factors necessary for transmission to occur are :

1. Contamination of the soil through direct defaecation or deposition of faeces on the ground; in other words poor sanitation.
2. Exposure of skin to infected soil, e.g. by going about bare-footed.
3. Suitable environmental conditions including :

(a) Eggs require warmth, shade, and moisture for development.

(b) Larvae require the following :

(i) optimum temperature 23-30°C.

(ii) oxygen, because their metabolism is aerobic, and
(iii) pH near neutrality.

4. Loose, humus soil with reasonable drainage and aeration.
5. Annual rainfall of 90 cm is necessary but must be spread uniformly throughout the year. *Nwosu* and *Anya* (1980) have investigated the effect of rainfall in Nigeria.

The following factors are unfavourable :

1. *Dryness*. Larvae are sensitive to desiccation, although they can survive on the film of water surrounding soil particles in otherwise dry soil.
2. *Low temperatures*. Less than 17°C will inhibit the development of eggs and frost is lethal to eggs and larvae.
3. Presence of undiluted faeces inhibits development and urine is lethal.
4. Direct sunlight.
5. Acid or alkaline soils.
6. Strong chemicals are lethal to larvae.
7. Salt in the soil inhibits hatching and is fatal to larvae.
8. Heavy clay, coarse, sandy soils and water-logged soil.
9. Alternate drying and moistening are damaging to the larvae, e.g. rainfall followed by long dry spells is unfavourable.

Other factors of epidemiological significance include :

1. Poverty and ignorance.
2. Shady locations as found in coffee, banana, and sugarcane plantations expose workers in such areas to greater risk of infection.
3. On the the other hand, workers in rice paddies are free of infections because of flooding.
4. People who prefer defaecation sites which are usually cool and shaded are exposed to continual reinfection when they repeatedly visit these sites.
5. Race, sex, and age are all factors of epidemiological importance. White people are ten times more susceptible than blacks and children more than adults and women more than men.
6. The nutritional status and iron reserve status of the population is very important.
7. Longevity of larvae; most can survive and remain infective for up to three weeks under ideal environmental conditions and even up to one year in protected areas like mines.

Nwosu (1981) has given an interesting account of the community ecology of soil-transmitted nematode infections of humans in Nigeria.

Revision Questions

1. Give an account of distribution and structure of *Ancylostoma duodenale*.
2. What disease is caused by *Ancylostoma* ? Describe its pathogenesis, therapy and prophylaxis.
3. Give an illustrated account of the life-history of *Ancylostoma.*
4. Write short notes on : (a) Rhabditiform larva (b) Filiform larva.

11

Wuchereria (=Filaria) bancrofti

The parasite gained importance after *Manson's* findings in 1878 that a mosquito acts as the transmitted agent for this parasite. *Manson* (1887) studied the life-cycle of the parasite and suggested reasons for the marked diurnal periodicity exhibited by filaria in the peripheral human circulation. *Vogel* (1928) and *Fain* (1951) gave accounts of the adult structure. The importance of this parasite to mankind is underline by the fact four International Meets were held in late 1960's to discuss the epidemiology, pathology and control of this worm.

SYSTEMATIC POSITION

Phylum	-	Aschelminthes
Class	-	Nematoda
Subclass	-	Phasmida
Order	-	Filaroidea
Genus	-	*Filaria*
Species	-	*bancrofti*

Distribution

Filaria, the most important human filarial parasite, is widely distributed in a tropical and subtropical circum-globular belt. But the parasite is not evenly distributed or uniformly prevalent throughout any country. It occurs almost entirely in costal areas and islands where there is a fairly long hot season with high humidity. In Africa found on the Mediterranean and east and west costal areas but not in the interior of Central Africa. In Asia it is prevalent on the coasts of Arabia, India, Malaya, Formosa, and North China and the southern part of Korea and Japan. It is prevalent in practically all the East Indian and South Pacific Island.

Habits and habitat

It is a *digentic parasite* i.e. its life-cycle is completed in two hosts.

The *definitive host* is man where it lives in the lymphatic system. The *intermediate host* is mosquito (*Culex, Aedes* and *Anopheles*).

MORPHOLOGY

The adult worms are found in the lymphatic vessels and lymph nodes of man only.

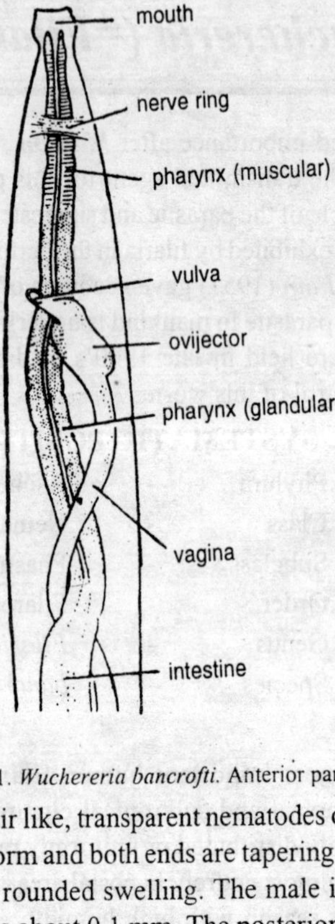

Fig. 11.1. *Wuchereria bancrofti.* Anterior part.

Adult worms are hair like, transparent nematodes often creamy white in colour. They are filiform and both ends are tapering. The posterior end terminates in a slightly rounded swelling. The male is about 40 mm, in length while thickness is about 0.1 mm. The posterior end of the male is curved ventrally and has two spicules of unequal length. Female measures 65-100 mm, in length and its thickness is about 0.2 to 0.3 mm. The tail end of the female is narrow unbent and abruptly pointed. Although females, liberate active embroys yet they are really *ovo-viviparous* i.e., lay egg with well developed embroys. Males and females usually remain coiled and can be separated with great difficulty. Females are more in numbers than males. The life span of the adult worm is about 4-5 years.

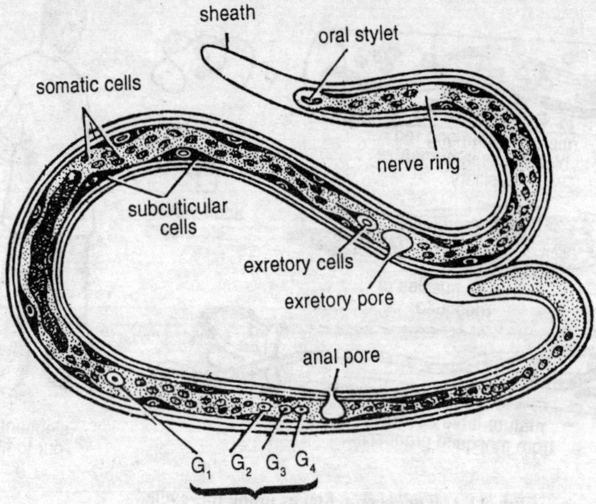

Fig. 11.2. *Wuchereria.* Microfilaria.

LIFE-CYCLE

Copulation takes place when individuals of both the sexes are present in the same lymph node. Female, releases numerous (about 1000) first stage larvae or juvenile called *microfilariae*. These larvae are surrounded by delicate membranes or sheaths and measure 225-300 mμ long and 10 mμ in diameter. When living they are colourless and transparent and may or may not be surrounded by sheaths. In order to identify them they are usually stained. The body wall contains a flattened epidermal cells and an inner column of cytoplasm containing nuclei. A nerve ring anteriorly, an excertory pore or 'V' spot, an excretory cell somewhat further back, a few genital cells posteriorly and an anal pore are present in body. The tail is devoid of nuclei.

The microfilariae appear in the peripheral circulation chiefly between 10 P.M. and 4 A.M. There has been much spaculation and experiment to determine the reason for this periodicity. According to one view, the larvae are concentrated in internal organs such as lungs (*Smyth*, 1976) during day time, but they circulate in the blood at night for transmission to the mosquitoes. Another view holds that embryos are born at a certain time each day and are then destroyed in the host within next 24 hours. According to *Manson-Bahr* and *Muggleton*, the nocturnal periodicity of microfilariae in peripheral blood of man is an adaptation to maintain a tryst with the night biting species of mosquito.

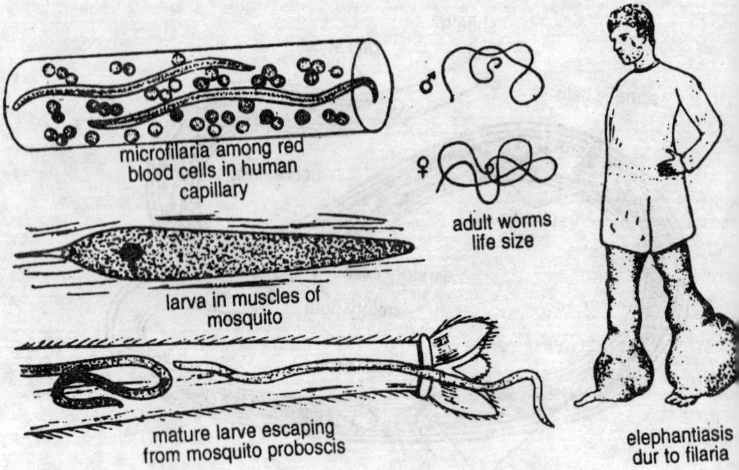

Fig. 11.3. *Wuchereria.* Stages in the life-cycle.

The further development of the microfilariae depend on their being sucked with blood by certain species of mosquitoes. In order to infect mosquitoes there must be about 15 or more microfilariae per drop of blood, a high concentration of 100 or more per drop of blood is fatal to mosquitoes.

Shortly after being digested by a mosquito the microfiariae penetrate through the stomach wall and migrate to the breast muscles. Here the embryos lie length wise between the muscle fibres. The size of body is reduced to half the original size length, but grows several times as thick. It now changes into a sausage-shaped creature.

The sausage shaped larva measures 1.5-2.0 mm. in length and 20-30 μ in diameter. As these changes take place the larva moults twice and becomes infective. The formation of the infective stage takes about 10-14 days and occurs best at 28°C and a relative humadity 90%. Digestive system, body cavity and genital organs also become well developed. This is called third stage larva. This infective stage reaches into the head of mosquito and finally into the proboscis on about 14th day. There are several larvae, coiled up in mosquito ready for infection in man. The complete development of microfilaria in the mosquito is from 10-20 days.

Infection and adult stage in man

When the infected mosquito bites a man, the third stage larvae are deposited, usually in pairs, on skin near the site of puncture. Later on the larvae enter through puncture or penetrate through the skin. From here they reach the lymphatic vessels and settle down at same spot, i.e., inguinal, scrotal or abdominal lymphatics. The larvae grow into adults. After a period of 5-18 months, larvae become sexually mature and start new generation of microfilariae.

Pathogenesis and clinical symptoms

The location of the adult worms in the lymph glands and connective tissues indicate that they would induce inflammatory and immune responses and, indeed, the pathogenesis is largely of this nature. Before patent microfilaraemia (i.e. before the microfilariae appear in the blood) the infection is symptomless. Then follows the acute stage marked by acute inflammation of various parts of the lymphatic system which become swollen and painful, accompanied by headache, nausea, urticaria, elephantoid or filarial fever with temperatures reaching 40°C (104°F), and sweating. The inflammation of the lymphatic system mainfests itself in lymphagitis, lymphadenitis (inflammation of the lymph ducts and lymph nodes respectively), lymphoedema, chyluria (lymph in urine making the urine milky in colour), lymph scrotum and hydrocoele, and lymph varies in which the lymph is blocked in the lymph ducts. If this conditions persists over a long period (usually for years) the lymphoedema becomes chronic, especially in the lower limbs, scrotum, vulva, breasts, and arms which become swollen and hard. There is fibrosis and hyperplasia of the tissues, and the skin thickens and shows folds and cracks in places, creating sites for secondary bacterial infection. Thus, the condition known as elephantiasis gradually develops.

It is now known that elephantiasis does not arise from blockage of the lymph ducts by adult worms, but is the result of allergic reactions to metabolic products of the worms over a long period. In fact, there are usually no microfilariae in the blood when elephantiasis sets in. In eastern Africa, particularly in Ethiopia, studies by *Oomen*, and *Price* have shown that in some areas where elephantiasis of the legs is endemic, *W. bancrofti* does not seem to occur, and so the condition can not be due to bancroftian filariasis. *Price* was able to show a link between the occurrence of elephantiasis and the type of alkaline volcanic soil common in Ethiopia. He, therefore, postulated that soluble silicates enter the body through wounds in the skin of the feet and legs, and crystallize in the lymph nodes, inducing inflammatory reaction which ultimately contributes to the onset of elephantiasis. Also, *Castellani* (1969) believes that elephantiasis is not directly caused by *W.bancrofti* since, even through filaricides may kill the worms, the recurring attacks of lymphagitis persists. He recognized many forms of elephantiasis and put foreard the 'Castellani theory' which shows that *Elephantiasis nostras* (found in the temperate zone) and *Elephantiasis tropica* (found in the tropics and supposed to be caused by filariae) are essentially the same disease whose etiology is bacterial (streptococcal). In the temperate zone, the bacterial infection is the primary cause while in the tropics it may be the secondary cause, the filaria merely serving as an irritating agent paving the way for bacterial infection.

Diagnosis

This can be done by finding microfilariae in stained thick blood film. It is sometimes necessary to distinguish different species of microfilariae in the blood. The presence or absence of a sheath, the location of internal nuclei and primordia of organs particularly in the posterior end, the relative position of the nerve ring, etc, are reliable features. *Gratama* (1970) has produced a scheme for differentiating *Microfilaria brncrofti* and *Microfilaria volvulus* in hydrocoele fluid in Liberia.

Haemagglutination and fluroescent antibody tests have proved successful for diagnosing chronic cases in which there is no microfilaraemio. Complement fixation and intradermal tests have not given reliable diagnosis.

Epidemiology

Some of the features of the epidemiology are sunmarized below :

1. The epidemiology of the disease depends on a complex of factors :
   (a) The distribution, local variations in density, susceptibility, longevity, and man-biting habits of the vectors.
   (b) The density, sleeping habits, and natural and acquired resistance of the human population.
   (c) Climatic factors such as high temperature and high relative humidity.

   So transmission should be seen as a dynamic system.
2. The vectors of urban and rural filariasis and their distribution and role in transmission in West and East Africa have been reviewed by *de Meillon* (1977) and treated above.
3. From all accounts, filariasis is increasing in prevalence and distribution in Africa, according to *de Meillon* :
   (a) The development of efficient waste disposal facilities has not kept pace with massive population migration to urban, suburban,

and industrial areas. The contaminated water in pit latrines, septic tanks, and drains, therefore, constitute good breeding habitats for *Culex fatigans* and hence increase in prevalence of urban filariasis.

(b) The spread of rural filariasis, on the other hand, has been enhanced by the development of dams, reservoirs, etc., which form ideal all-season habitats for the vector *A.gambiae.*

4. Filariasis is rare among children. The incidence increases with age, but it is observed that those who show micrifilaraemia have been exposed to repeated bits of infected mosquitoes over a long period.
5. There are over 250 million cases of filariasis in the world, over half of which are in South-East Asia, particularly India. Because the prevalence of the disease is higher in East Africa than in West Africa, it has been proposed that the disease was introduced into Africa from South-East Asia via East Africa.
6. However, *Wijers* (1977) suggests that bancroftian filariasis in Africa and India may be caused by two different strains of the worm, because of differences in the parts of the body affected. In East Africa, hydrocoele is a common feature of the disease, over 50 per cent of infected males being affected. In India, on the other hand, elephantiasis of the legs predominates over hydrocoele. In West Africa, though high microfilarial rates (40 to 50 per cent) may be encountered, there are only a few clinical manifestations of the disease, suggesting the possibility of a third strain.
7. The absence of microfilariae in chronic cases suggests the development of immunity. Thus reliance on blood smears for microfilarial rates alone may give a misleading picture of the epidemiology.
8. Also, misidentification of microfilariae can give a misleading idea of the epidemiology. For example, *Nelson*, *Hsieh*, and *Furlong* (1962) found microfilariae of six different speices of animal parasites in the mosquito *Aedes pembaensis* which was previously thought to be the vector of *W. bancrofti* in Kenya. Similarly important is the accurate identification of vectors, but there is no international reference centre for mosquitoes.
9. Accounts of the epidemiology of bancroftian filariasis and transmission by mosquitoes have been given in various publications

by *Brengues* and *Subra* and their co-workers and *Diallo* and colleagues in West Africa and *Wijers* in East Africa (already cited), as well as by *Kuhlow* and colleagues in Liberia and *McConnell*, *Ashafa*, and *Dennis* (1976) in Ethiopia.

Prevention and Control

Control follows the general programme for mosquito control treated in greater detail later under malaria control. The prospect is, however, gloomy because of resistance to insecticides. Larviciding of pit latrines and drains may give good results, but better still is the provision of adequate sewage disposal facilites which unfortunately is beyound the means of countries in endemic areas, mainly because of the prohibitive costs. Control of rural filariasis is much more difficult. Mass chemotherapy of the population is an attractive proposition but fraught with all kinds of problems, seemingly insurmountable. In short, 'up to the present date, effective measures of control of bancroftian filariasis have not been practical in Africa'.

Prevention of mosquito bites by the use of mosquito nets and insect repellants can also help. This could reduce transmission to such low levels that even through microfilariae might still be present, the more severe form of hydrocoel and elephantiasis will hopefully become a rarity.

Treatment

Drugs for *filaricidal* reactions can be grouped into 3 main groups:

1. On adult worms - Mel w, an arsenical preparation.
2. On microfilariae - Diethycarbamazine (Hetrazan).
3. On infective larvae and immature adult worm para-melaminyl-phenylstibonate (MSb)

Prophylaxis

It includes :

1. Killing the adult intermediate host i.e., mosquitoes.
2. Use of selected insecticides to kill the adult and larval stages of mosquitoes.
3. Use of wire netting at door and windows of human dwelling to prevent the entry of mosquitoes.
4. Administration of diethylcarbamagine (DEC) mixed in common salt has shown encouraging results.

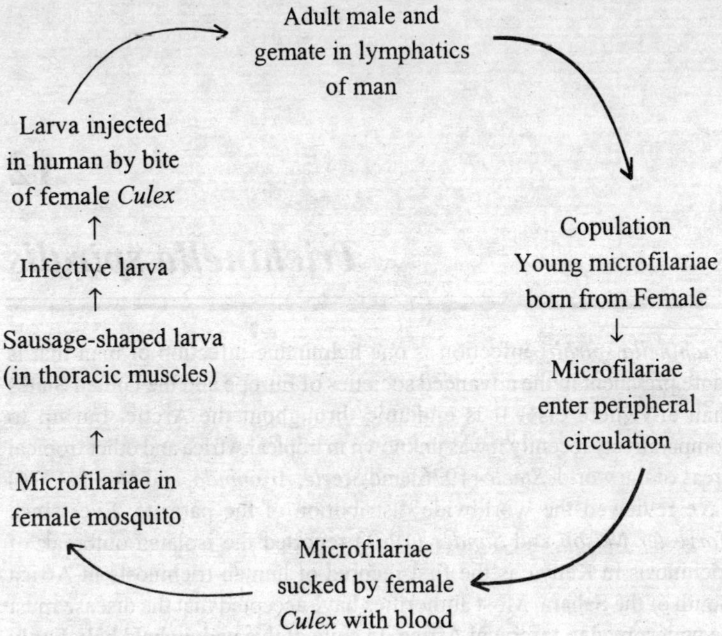

Life cycle of Wuchereria.

Revision Questions

1. Describe the distribution and structure of *Wuchereria.*
2. Describe the structure and life-history of any nematode studied by you.
3. Write short notes on microfilaria larva.

12

Trichinella spiralis

Trichinella spiralis infection is one helminthic infection of man that is more prevalent in the advanced societies of Europe and the United States than anywhere else. It is endemic throughout the Arctic, but up to comparatively recently it was unknown in tropical Africa and other tropical areas of the world. *Steele* (1976) and *Steele*, *Arambulo*, and *Moon* (1976) have reviewed the worldwide distribution of the parasite. Ever since *Forrester Nelson* and *Sander* (1961) reported the isolated outbreak of trichinosis in Kenya as the first recored of human trichinosis in Africa South of the Sahara. Most authorities have accepted that the disease must be widespread in toropical Africa. In spite of this widely held belief only a few cases of human trichinosis have uptill now been reported from tropical Africa. In fact it is fair to say that the information to go on derives

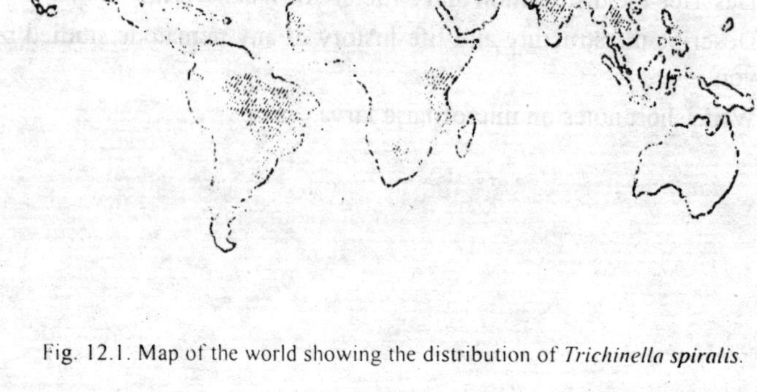

Fig. 12.1. Map of the world showing the distribution of *Trichinella spiralis*.

from reports of isolated and sporadic outbreaks mainly one in Kenya one in Tanzania and one from dakar Sanegal. Some of the reasons adduced for this include pooor diagnostic facilities difficulty of diagnosis and possibility of overlooking of misdiagnosing the disease because its symptoms are similar to those of other tropical infections which is all very well. But medical services have inproved and a service which has been able to diagnose such a wide array of parasitic infections throughout the continent could not have failed to throw up trichinosis if it were anything but the disease of extreme rarity that it is. The incidence of trichinosis in wild animals in tropical Africa is, however, another matter because high provalences have been reported in surveys in Kenya by *Nelson Guggisberg*, and *Mukundi* (1963), in Tanzania by *Sachs* (1970), and in Senegal by *Gretillat* (1970), *Gretillat* and *Chevalier* (1970), and *Gretillat* and *Vassiliades* (1967).

Trichinella is the only genus of the family Trichinellidae of the order Trichurate and for a long time it was considered to be monospecific consiting of the single species *T.spiralis* many strains or biological races of this species have been recognized differing in their degree of infectivity in various experimental hosts their pathogenicity and in other physiological features though they are morphologically indistinguishable. Many authors recoginze these merely as strains, while others have eleveted them to specific status: *T. spiralis* for the north temperate strain *T. nelsoni* for the Tropical and *T. nativa* for the Arctic strains and even a fourth species *T. pseudospiralis*. The controversy on the speciation of *Trichinella* is very interesting, but beyond the scop of this book. *Sukhdeo* and *Meerovitch* (1977) have presented fresh experimental evidence which tends to support the existence of three species suggesting that they are incipient species on the way to becoming completely separated as a result of their geographical and hostal isolation *Madsen* (1976), on the other hand in a most interesting review shows that there in insufficient scientific evidences to justify the creation of new species whose desciptions do not in any respect stand up to the international code.

MORPHOLOGY

1. *Trichinella spiralis* showa most of the features of the order Trichurate- stichocytes, lack of spicules in males, single female gonad terminal, anus, etc.
2. Size : It is the smallest human nematode parasite males measuring 1.4 to 1.6 mm and females 3.0 to 4.0 mm in length.
3. The whip-like appearance characteristic of the order is not so marked, the body merely tapering towards the anterior end.

4. Male has two large papillae.
5. Females are oviviparous : the posterior part of the uterus is filled with developing eggs, while the anterior portion contains hatched larvae.

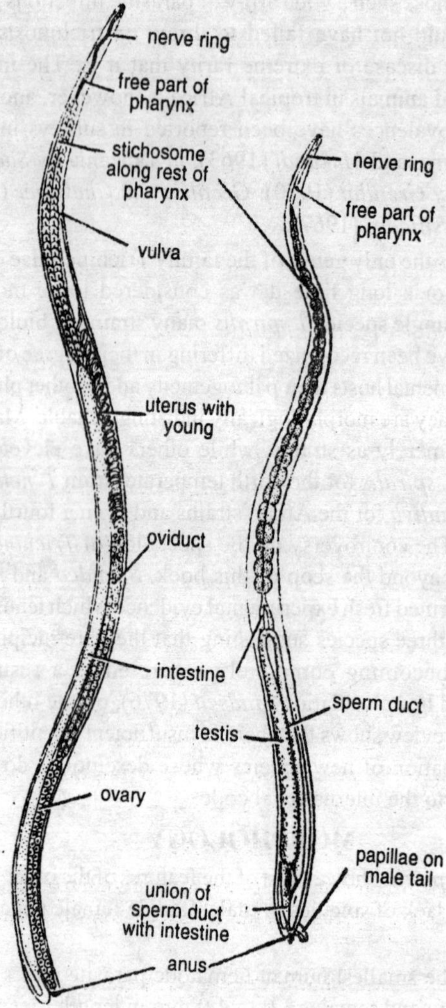

Fig. 12.2. *Trichinella spiralis*. A—Male, B—Female.

LIFE CYCLE

There is one remarkable feature of the life cycle of *T. spiralis*; the larvae and adults develop in different organs or one animal i.e., a single animal serves as both intermediate and definitive host. The adults live in the intestine of man and other suitable hosts. After copulation, the females burrow into the mucosa and submucosa where, being ovoviviparous they produce live larvae (about 1500 per female) These larvae then pantrate further finally reching the blood stream which distributes them to various organs of the body. They tend to, settle, however, in skeletal muscles

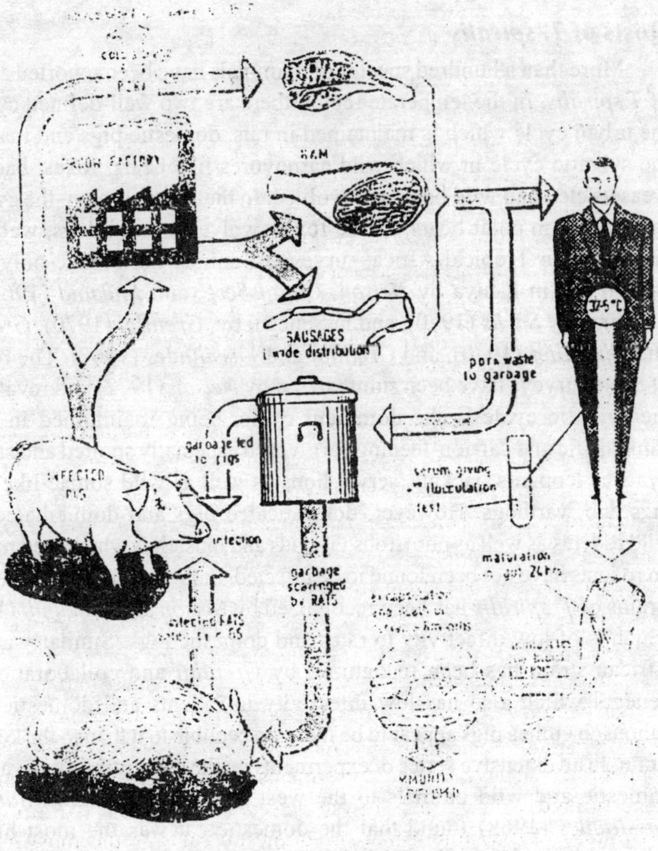

Fig. 12.3. *Trichinella spiralis.* Life cycle and probable mode of transmission in Europe and American communities.

especially muscles of the diaphragm, jaws, tongue, etc. Here they penetrate individual muscle fibres and grow up to ten time their orginal size reaching up to 1.0 mm in eight weeks during this period they begining to coil forming the characterstic spiral. Host tissue reactions form a lemon-shaped cyst around the larva which is now known as the encapsulated larva within this cyst the larva can survive for several months or years and still be viable. However, after about 6 to 18 months, calcification of the cyst and later of the larvae themselves sets in .Infection occurs when infected muscle is consumed by a suitable host. The larvae are released in the duodenum and develop rapidly reaching sexual maturity in only 40 hours. *Shanta* and *Meerovitch* (1967) have given good accounts of the intestinal and muscle phases of the life cycle.

Hosts of *T. spiralis*

More than a hundred species of mammals have been reported as host of *T.spiralis*. In the temperate region there are two well-defined cycles : the urban cycle which is maintained in rats, domestic pigs and man and the sylvatic cycle in which wild carnovores like bears, foxes, badgers, weasels etc., and wild boar are involved. In the Arctic region, the cycle is maintained in polar bears, Arctic foxes, wolves, minks, dogs, walruses, and seals, In Tropical Africa, surveys of animal hosts have only been carried out in Kenya by *Nelson*, *Guggisberg*, and *Mukund* (1963), in Tanzania by *Sachs* (1970), and in Senegal by, *Gretillat* (1970), *Gretillat* and *Chevalier* (1970), and *Gretillat* and *Vassiliades* (1967). The results of these surveys have been summarized by *Nelson* (1972) and revals that the sylvatic cycle is the dominant cycle, being maintained in large canibalistic and carrion-feeding carnivored especially spotted and striped hyaenas leopards, jackals, serval lions as well as wild suidae like bush pigs and warthogs. However, domesticated pigs and domiciliated and wild rodents as well as numerous viverids and mustelids which feed mainly on rodents have not been found to be infected. In this connection, a Kenyan strains of *T. spiralis* has been recognized by *Nelson* and *Mukundi* (1963) which is of low infectivity to rates and domestic pigs. Similarly a west African strain has been recognized by *Gretillat* and collaboratiors in Senegal which also has low infectivity in rodents and domestic pigs althousgh guinea pigs appear to be more susceptible to it than to the kenyan strain. In an extensive series of experments on the susceptibility of various domestic and wild animals to the west African strain. *Gretillat* and *Vassiliades* (1968) found that the domestic cat was the most highly susceptible closely followed by the domestic dog. The domestic had very low susceptibility, while rats, mice gambian rates, hedgehogs, and rabbits were refractive to infection. The primates, *Papio papio* and *Eurythocebus patas* ware very susceptible while the warthog, which is regarded as and occasional host was difficult to infect experimentally. Finally carnivores proved to be the preferred hosts in west Africa. There is apparently an intestinal barrier in the pig which prevents most larvae from developing to adulthood while a muscular barrier causes rapid calcification of 60 to 100 per cent of the larval cyst. Finallly, and most inportantly it has been established that serial passage of this train from wild carnivores through white rats increased its infectivity and rendered it capable of infecting domestic pigs. This is of great epidemiological significance because pigs can readily become susceptible when exposed to larvae which have undergone repeated passage, a fact stressed by *Madsen* (1976) in connection with the validity of strains.

Pathogenesis and pathology

The symptoms of trichinosis are characterzed by vagueness, and mimic symptoms of other conditions such that over fifty diseases have been diagnosed for it. Details of the pathogenesis and pathology in tropical Africa have been given in the case reports of the outbreaks by *Bura* and *Willett* (1977) and *Forrester*, *Nelson* and *Sander* (1961). According to *Kagan* (1978), of some seventy- five symptoms and signs associated with trichinosis, the most impotant in order of frequency of occurrence are : fever, gastro-interstinal symptoms including diarrhoea, muscle pain, nausea, oedema of the face and eyelids, eosinophilia, rash, weakness, and neurological symptoms. Death, when it occurs, is the result of heart failure, respiratiory complications, toxemia or kidney malfunction. All these symptoms are the results of the effects of adult worms in the intestinal mucosa, larval migration and their penetration and encapsulation in the muscle fibres.

One important feature of the East Africans cases is the density of larvae per gram of muscle tissue which is the highest ever recorded for man. In Kenya larval density ranged from 1850 to 5190 while in Tanzania up to 6000 cysts per gram was reported, with an average of 3390. These figures exceed, by a wide margin the findings in Liverpool where densities of 7.5 and 119 larvae per gram were classified as severe or very severe respectively. The fact that these densities produced conditions of equal clinical severity suggests that the East African strain is of higher infectivity and lower pathogencity than the strain in Liverpool.

Diagnosis and treatment

Disgnosis of trichinosis presents a number of difficulties :

1. Eggs or larvae are usually not detected in faecal samples.
2. Light and moderate infections show no symptoms.
3. Even heavy infections present vague symptoms that are usually cofused with other conditions. Thus trichinosis may be overlooked or missdiagnosed if it is not supected in the first place.

There are a number of diagnostic techniques for detecting encapsulated larvae in the musscles :

1. Digesting a biopsy specimen of muscle in artificial gastric enzymes such as pepsin and dilute hydrochloric acid overnight, and recovering the lervae by sedimentation in a tube.
2. Using a trichinoscope to compress a muscle sample between glass plates, such that it appears semitransparent under the light microscope revealing the larvae.

Unfortunately, larvae in muscle biopsy specimens in cases of early and light infections are not easy to find. Furthermore, selection of sites for diagnostic biopsy depends on knowledge of predilection sites of larvae in the muscle of the body, but not many detailed studies of this have been reported. However, it is generally assumed that, as in rats and pigs, the diaphragm is the preferred site followed by the jaws, tongue, larynx, and eye. Biopsy speciments are, however, obtained from the gastroneminus or deltoid muscle.

Other tests include

3. Immunodiagnostic techniques : flocculation tests and interdermal tests, although positive reaction with the latter does not necessarily mean the existence of current infection.
4. Xenodiagnosis, in which muscle biospsies are fed to laboratory rates and the presence of infection is confirmed if larvae develop in the muscles of the rat.
5. The presence of peri-orbital oedema especially when it occurs in a group of patients, is a useful diagnostic clue.

There is no satisfactory drug for specific treatment of trichinosis. Thiabendazole has been shown to kill encapsulated larvae and adult stages in experimental mice but results of clinical tests with man are equivocal At present, efforts are directed towords relieving the symytoms, and for this corticosterodis and analgesics come in handy. A review of chemotherapy of trichinosis has been given by *Campbell* and *Blair* (1974).

Epidemiology and Control

The literature on the epidemiology of trichinosis in the temperate zones of Europe and North America is extensive.

As stated above, there are two maintenance cycles : the sylvatic cycle involving wild carnivores and their prey or carrion and the urban or domestic cycle involving the man-domestic pig-rat traingle'. The situation is graphically represented by *Nelson* (1972). In this region, the urban cycle plays the dominant role in epidemiology, man becoming infected by consuming undercooked pork and infected pork products like sausages. Rats become infected by practising cannibalism or by feeding on rat faeces containging infective larvae. Pigs themselves maintain the cycle by being fed garbage containing infected pork scrap and offal or by eating each others tails especially in overcrowded piggeries. It is also believed that pigs can gain infection by eating rats in their sties. According to *Madsen* (1976) this 'predation theory' is responsible for the lack of proper appreciation of the significant role of the encapsulated larva. The evidence has been piling up since the last century that the encapsulated larva is in effect a free-living stage, surviving in the muscle fibres of the decomposing carcass (carrion), a situation analogous to the free- living eggs or larvae of other nematodes. This evidence according to *Madsen* has been largely ignored because of the dominance of the predation theory. Afresh look at the domestic cycle will indicate that what happens is that pigs gain infection by eating dead and decomposing carcasses of rats, while rats gain access to decomposing scraps of infected pork.

In tropical Africa, the domesitc cycle is of little or no significance, mainly because, as stated above, the domestic pig and rats have low susceptibility to the African strain of the parasite. In this respect, the epidemiology is remarkably similar to that of the Arctic region. In tropical Africa therefore, the sylvatic cycle is the dominant one, the cycle being maintained in carnivores. In Eest Africa, it is significant that the hyaena is the commonest host with a 44 per cent infection rate. It is significant because the hyaena is a scavenger thriving mostly on carrion and occasionallly practising cannibalism as well. According to *Madsen* (1976), wild carnivores consume more carrion than is generally realized thereby reinforcing the belief in the inpotance of putrefaction of infected meat in the epidemiology of trichinosis. With the carniovres feeding on carrion or on flesh of other carnivores and even practising cannibalism it can be seen that wild rodents are not essential for the cycle in Africa. However, wild suids like bushpigs and warthogs come into the cycle again either through carrion-feeding or cannibalism.

Since people in tropical Africa, as in most other parts of the world, do not relish the flesh of carnivores, transmission is most likely to occur by eating the undercooked flesh of wild suids. Indeed, the three well-documented cases from Kenya, Tanzania and Senegal involved the consumption of the flesh of the bushipig and warthog. The Senegal case is interesting because it involved nine Europeans in Dakar who indulged themselves by feasting on barbecued worthog, an obvious act of indiscretion in the tropics, or anywhere else for that matter !

With regard to the outberak in Kenya reported by *Forrester*, *Nelson*, and *Sander* (1961), *Nelson* (1972) attributes it to change in human behaviour brought about by the mau mau insurrection which made young kenyans abandon old Kikuyu tradition of disdain for the flesh of wild animals, and eat undercooked meat to bushpig. However, this explanation may not be wholly acceptable because in the first place, people do not require an inssurrection (a struggle for political independence in this case) in order to change or discard traditional practices. In the second place, wild pig in commonly consumed in different parts of Africa where trichinosis is not a serious problem.

The absence of trichinosis form most parts of tropical Africa has intrigued most workers in the industrialized countries of the temperate zones. In their search of a scientific explanation of this phenomemon a veriety of opininos have been expressed. For example *Nelson* (1972), *Nelson Guggisbrg*, and *Mukundi* (1963), and *Pawlowski* (1974) believe that *T.spiralis* infection is widespread and will come to light if we search hard enough. Another view is that trichinosis has been missed so far simply because of its poor diagnostic features or because it presents symptoms which simulate other more common and familiar tropical diseases, so that the suggestive symptoms are overlooked if trichinosis is not suspected in the first place. Finally, it has been predicted that there is the possibility of changing epidemiology either through the introduction of infective strians with the importation of exotic breeds of pigs in the expanding pig indiustry or as a result of repetitious transmission through rats and pigs which will increase the infectivity of the strain. However, a simple explanation may be that this is an infection which endemic in the temperate region but sporadic in tropical Africa.

In conclusion the essential feature of the epidemiology of trichinosis in tropical Africa is that sylvatic trichinosis is prevalent and the cycle is maintained in wild cannibalistic, or carrion-feeding carnivores and wild suids. Man, however, occasionallly become involved by eating undercooked flesh of wild suids and this results in outbreaks of the disease in isolated part of the continent.

Prevention and control

The obvious preventive measure is by eating well-cooked pork. Control measures include cooking pork at a temperature of 55°C or 15 minutes per pound or alternatively freezing pork products at—40° C for 30 minutes, —25°C for 24 hours, or at - 15°C for 6 days (*Smyth*, 1976). Such treatments are adequate to kill the encapsulated larvae. Also, drying or salting pork for long perods renders it safe for human consumption.

It is now generally accepted that the prevalence of trichinosis is declinging throughout the world ; indeed, it was never of great importance in the first place. Reduction is transmission in the United States was achieved as a result of legislation against feeding undercooked or raw garbage to pigs. However, the irony, according to *Nelson* (1972), was that this legislation was not directed primarily against the human disease, but rather against vesicular exanthema, a virus infection of swine of great veterinary and econmic importance to the piggery industry.

13

Reproduction in Nematodes

In group Nematoda sexes are separate and sexual dimorphism is well distinct, females and males can be distinuished by their primary and secondary sexual characters. Males can be readily separated from females by the presence of a copulatory apparatus and in many cases by their smallar size. In some animal parasites, the female is 30 times larger than the male or the male may be so small that it enters the female and lives within the uterus. In some palnt-parasitic genera like *Heterodera* and *Meloidogyne* the two sexes show a pronounced dimorphism. The female of these genera enlarges to become a reproductive sac, whereas the male remains slender and worm-like. In other nematodes sexual dmorphism may be also attrributed to failure of the males to attain complete development. Thus in males of genera *Paratylenchus*, *Hemicycliophora* and *Criconemoides*, the stylet and oesophagus are often not as well developed as in the females.

REPRODUCTIVE SYSTEM

The reproductive system is similar in both sexes of all nematodes. It is composed of 1 or 2 tubular gonads which may vary greatly in length and may be straight, reflexed or coiled. Females with 2 complete genital tubes are called *didelphic* whereas females with 1 gential tube are called *monodelphic*. The correspoding terms in the male are *diorchic* indicating the possession of 2 testes and *monorchic* indicating the possession of 1 testis. If the uteri are opposed at origin the females are termed *amphidelphic*, if they are parallel and anteriorly directed at origin, the females are called *prodelphic* and if the uteri are parallel and directed posteriorly at origin the females are called *opisthodelphic*.

Female Reproductive System

In most nematodes the germ cells are proliferated only in the proximal end of the gonad. This type of germ cell production is termed as *telogonic*. A relatively small number of nematodes possess hologonic gonads in which the germ cells are formed along the entire length of the gonad.

The Ovary

The Proxinal end of the telogonic gonad in females is occupied by the ovary, a tubular sac in which the germinal cells develop. This sac is made up of an epithelial layer and a germinal cord. In general, the overy can be subdivided into two regions : (1) The *germinal zone* : a region in which rapid division of cells takes place ; (2) The *growth zone* : a region of gradual increase in size of the oogonia. The oogonia are attached to the rachis which is a central strand of non-nucleated tissue. The function of the rachis is uncertain. In some nematodes, the oogonia are arranged around the rachis in the growth zone of the ovary forming a cylinder. The growth zone in most free-living nematodes consists of a single chain of cells and the rachis disappears when the oogonia assume the single file. The free end of the ovary may be either reflexed or it may be outstretched.

The Oviduct

The ovary is followed by the oviduct. The oviduct is a functional rather than a structural entity of the female gonad. It is a constricted, thick-walled region between uterus and ovary and consists of a narrow tube with a high columnar epithelium. In some nematodes, a slight enlargement in the proximal part of the oviduct serves as a *spermatheca* or *seminal receptacle*.

Uterus

The oviduct is attached to the uterus which is a broad tube lined with a flat cuboidal epithelium and covered by a muscle layer of circular and oblique fiberes. The beginning of the uterus usually functions as a spermatheca and fertilization occours in this region. The eggs are stored in the uterus and as they pass distally along it, they undergo fertilization and matuation, shell formation and some degree of embryonic developmant.

Monodelphic nematodes generally possess a postvulvar uterine sac which may function as a spermatheca. Such a uterine sac may be the remainder of a second genital tube.

Distally the uterus or uteri enter a common tube, the *vagina* which is lined with cuticle and provided with muscles. The vagina is usually quite short and opens through the female *gonopore*, the *vulva*.

The vulva is a midventral transverse slit that is usually situated along the middle third of the body but may be shifited anterior or posterior to the vicinity of the anus. In most cases, a specialized musculature is present which dilates the vulva.

If free-living Adenophorea are compared with free-living Secernentea, a tendency may be noted toword the formation of smaller, but more numerous eggs in Secernentea. This is often accompanied by a lengthening of the growth zone of the ovary.

Parasitic nematodes are usually characterized by increase in the size of the entire genital tube. The uteri and the growth zone of the ovary are especially lengthened and muscle development here is greater. This is correlated with higher egg production.

Male Reproductive System

In most of nematode the testes are of monorchic type and the male reproductive system has the following organs :

1. Testis

The testis of telogonic nematodes can also be subdivide into a *germinal zone* and a *growth zone*. The maturation of the sperms takes place at the end of the growth zone. The sperms are of different shape in the various nematodes. They may be amoeboid or rounded as in many Secernentea, or they may be flagellate or conical. The free end of the testis is often reflexed and long testes may be curved several times.

The testis is followed by the *seminal vesicle* which is a dilated portion of the male gonoduct and functions as a storage for sperm. A few nematodes possess a *vas efferens*. This is a specialized tubular part of the seminal vesicle which separates the growth zone of the testis from the seminal vesicle.

2. Sperm duct or Vas deferens

The main part of the male gonoduct is the vas deferens. It is usually composed of tubular and glandular regions. It may be provided with musculature throughout its length or only in its terminal portion. In some nematodes, the vas deferens is differentiated into an anterior glandular region comprising the greater part of its length, and a posterior non-glandular region or ejaculatory duct. In *Rhabditis strongyloides*, the vas deferens has 2 larger lateral pouches which extend anteriorly. These pouches are termed *ejaculatory glands* and probably secrete the adhesive cement deposited on the vulva of the female during copulation.

Nematodes of the class Secernentea characteristically have a single testis. The only exception to this is the genus *Meloidogyne*. These nematodes possess either 1 testis or 2 testis which proceed posteriorly and unite in a single vas deferens. Most Adenophorea are diorchic. Typically, the 2 testes extend in opposite directions but are joined to form the vas deferens as a single tubular canal directer posteriorly.

In males of parasitic nematodes the reproductive system is increased in length as compared to males of non-parasitic nematodes.

MODE OF REPRODUCTION

Bisexual species usually have equal numbers of females and males. Insemination by a male and egg fertilization in necessary for the development of the eggs in such forms. In many free-living and plant-parasitic nematodes, however, males are much fewer in number than

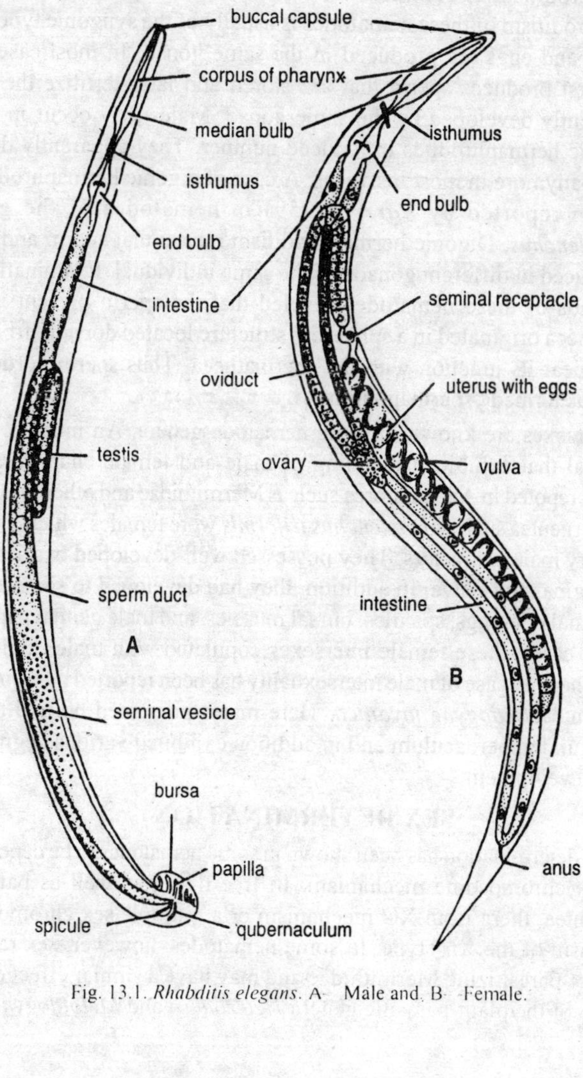

Fig. 13.1. *Rhabditis elegans*. A- Male and. B- Female.

females. This scarcity of absence of males indicates a tendancy towards hermaphroditism or parthenogenesis. From the bisexual condition with equal numbers of males and females, all intermediate types to hermaphorditic and parthenogenetic species can be found, wherein males have become rare individuals and have mostly lost their sexual instinct, or wherein males may be completely lacking.

In hermaphroditic species, reproduction through self fertilization occours. Hermaphroditic nematodes have the appearance of females. Hermaphroditism is common among the free-living Rhabditida. The hermaphroditism of these nematodes is usually of the syngonic type, that is sperm and eggs are produced in the same gonad. In most cases the gonad first produces sperm that are stored and later fertilize the eggs subsequently developed by the same gonad. Males may occur in these protandric hermaphrodites in reduced number. They apparently do not function anymore in most instances. A case of digonic hermaphroditism has been reported by *Parry* (1959) in nematodes of the genus *Helicotylenchus*. Digonic hermaphroditism means that sperm and eggs are produced in different gonads of the same individual. Examination of the gonads of these nematodes reveled that the sperm present in the spermatheca originated in a sphaeroid structure located dorsally from the oviduct near its junction with the spermatheca. Thus sperm porducing organ was termed "spermagonium".

Intersexes are known in some nematode genera. An intersex in an individual that exhibits a blending of male and female characters. Al intersxes repoted in Adenophorea such as Mermithidae and other Enoplida and Secernentea such as *Ditylenchus triformis* were females which showed secondary male characters. They possessed well-developed ovaries with uteri, vagina, and vulva. In addition, they had developed to some extent male sexual openings, spicules, bursal muscles, and male genital papillae In many cases, these female intersexes copulated with males and were fertile. The only case of male intersexuality has been reported by *Chitwood* (1949) in *Meloidogyne javanica.* Here males possessed normal testes spicules und gubernaculum and in addition exhibited various degrees of vulvar development

SEX DETERMINATION

Sex determination has been shown in some nematodes to be dependent on a sex chromosome mechanism. In free-living as well as parasitic Secernentea, there is an XO mechanism or a multiple sex chromosome mechanism of the XnO type. In some nematodes, however, sex ratio in the insect-parasitizing Mermithidae and may have a similar effect on the sex ratio of the plant-parasitic genera *Heterodera* and *Meloidogyne*.

REPRODUCTIVE HABIT

Little in known about the reproductive habits of nematodes. Some aquatic and terrestrial nematodes lay a small number of relatively large eggs, 50 or less per female, over a long period of time. In saprophagous forms such as *Rhabditis, Diplogaster*, *Panagrolaimus* a single female may lay more than 200 eggs within one week. Plant-parasitic nematodes, if they are endoparasitic, deposit their eggs inside the plant tissues. Animal-parasitic nematodes discharge their eggs inside the host body. Parasitic nematodes lay more eggs per female than free-living forms.

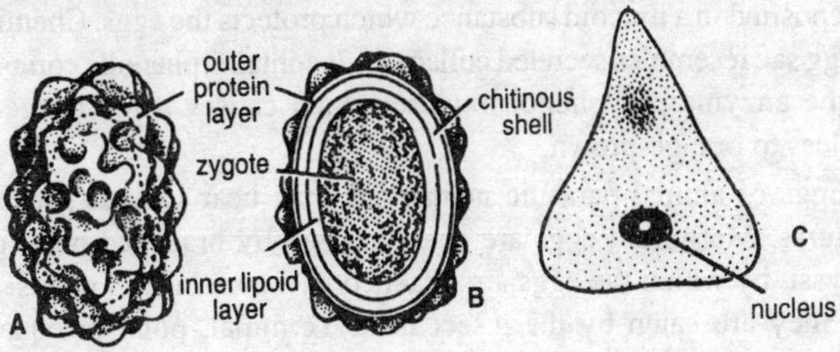

Fig. 13.2. *Ascaris,* A—Entire egg, B—Egg in section, C—A sperm.

The female may lay unsegmented eggs or they may undergo segmentation in the uterus. If the eggs are discharged unsegmented or in various degrees of development, the condition is termed *oviparous*. In many of the animal-parasitic nematodes, the eggs develop completely while in the uterus and the larvae hatch therein. This is called ovoviviparity. In certain Rhabditida, a peculiar phenomenon has been observed. Older females are unable to discharge the eggs. The larvae thus hatch in the uterus and after the death of female, break through the uterine wall into the coelom where they soon consume the body contents of the female. The larvae remain enclosed in the cuticle of the mother individual for a period of time until they escape and continue their dvelomment normally. This phenomenon has been described by *Seurat* (1920) as "Endotokia matricida" and has been mentioned later by other investigators.

NEMATODE EGGS

Eggs of most nematodes are covered by 3 distinct membranes : (1) an external protein coat which is a secretion of the uterine wall; (2) a chitinous membrane of true shell which secreted by the egg itself; and (3) a vitelline membrane beneath the chitinous membrane. These 3 layers are not present in all nematodes. In some groups, such as the Rhabditoidea and Tylenchida, the protein coat is absent in most of the species. In general, the morphology of the egg varies according to the life habit of the nematode concerned. Usually the eggs have a rounded, oval or elliptical shape.

In some aquatic nematodes, such a *Anaplectus* and *Trilobus*, the eggs have hooks, spines, or small excrescences which are formations of the protein layer.

Among the plant parasites of the order Tylenchida, there is considerable variation in the relative length of the egg. The eggs, in general, have a smooth surface. In species of *Meloidogyne* and *Heterodera*, they are deposited in a mucoid substance which protects the eggs. Chemically, this egg sac resembles secreted collagens. It contains phenolic compounds and the enzyme polyphenol oxidase which causes it to change from colorless to orange-brown.

Eggs of animal-parasitic nematodes may bear various specialized structures. Mermithid eggs are characterized by branched polar cords, the byssi, by which the eggs are attached to the surface of plants. From here, they are eaten by the insect host. Terminal, polar or equatorial filaments may be present in other animal-parasitic species. These filaments may function in a similar manner as the byssi or may serve to hold the eggs together in clusters and thus insure better infection of the hosts. The egg shell may be mammillated or may exhibit opercula. Opercula are slightly sclerotized zones in the egg membranes. They are usually located at the egg poles and make it easier for the larvae to escape from these thick-shelled eggs. Mammillation refers to the rounded excrescences over the surface of the protein coat as exhibited by the eggs of *Ascaris lumbricoides*.

DEVELOPMENT

Oogenesis

Oogenesis has been studied in free- living, animal and plant-parasitic nematodes. The same is true for nematode embryology which has been the subject of several classical studies in cytology.

Parascaris equorum bivalens, a large animal parasite, shall be selected here to review the maturation cycle and embryology of a bisexual nematode species.

After insemination by male has taken place, the sperms travel up the female reproductive tract and fertilize the ripe egg cells as they proceed distally. As soon as one sperm has entered the egg cell, a fertilization membrane appears around the egg and the egg protoplasm contracts. The shell begins to form by an endogenous process immediately after sperm penetration. A thin lipoid membrane, the vitelline membrane, is produced on the inner side of the shall. As the egg passes along the uterus the protein membrane is added outside the shell. While the egg membranes are being formed, 2 maturation divisions occur by means of which the normal diploid female chromosome set of the egg cell is reduced to a single or haploid set. *Parascaris equorum* bivalens has 2 pairs of chromosomes (2n=4). After the sperm has entered the egg, the first maturation division of the egg nucleus initiates. Early in the prophase the homologous chromosomes pair and then each chromosome is doubled, thus two tetrads are formed. The tetrads take the usual position at the equatorial plate of the cell and the first division of the egg nucleus follows. Two chromatids of each tetrad get into the first polar body which separates from the egg. The 2 chromatids of each tetrad that remain in the egg rotate and the second division follows. The second polar body is formed and the egg nucleus contains now the haploid number of chromosomes (n=2). During the second maturation division, the sperm nucleus increases its volume and the chromosomes appear as in a resting nucleus. The same change takes place with the egg nucleus soon after its second maturation division is completed. These nuclei are the male and female pronuclei. Without fusing each pronucleus again develops 2 chromosomes which immediately, take their position at the equatorial plate of the cell. The first cleavage spindle in formed. Each female and male chromosome doubles and the first cleavage division takes place. This is a mitotic division.

Embryonic Development

The first cleavage occurs transverse to the egg axis and produces 2 cells of about equal size. The second cleavage results in 4 cells which are at first arranged in a "T" shape. but later shift to the rhomboid shape characteristic of the four-cell stage of nematodes. Subsequent mitotic cleavages result in a coeloblastula which is generally dorso-ventrally flattened and elongated in the antero-posterior direction. The coeloblastula is followed by the gastrula stage. Gastrulation usually occurs by epiboly. The ectodermal sides of the embryo grow down and push the ventral surface into the interior. The invaginated material forms an oval mass and extends anterio-posteriorly in the blastocoel. By the process of gastrulation, the previously flattened embryo is altered to a cylindrical shape. At this stage, the 3 germ layers of the nematode are clearly defined. In cross section the embryo consists of an ectodermal epithelium enclosing a central column consisting of endoderm in the middle with a strand of mesoderm to either side. The embryo then continuous to grow and is soon recognized as a larva inside the egg membranes.

The embryonic development of nematodes has a highly determinate character. The fate of each cell can be definitely followed during

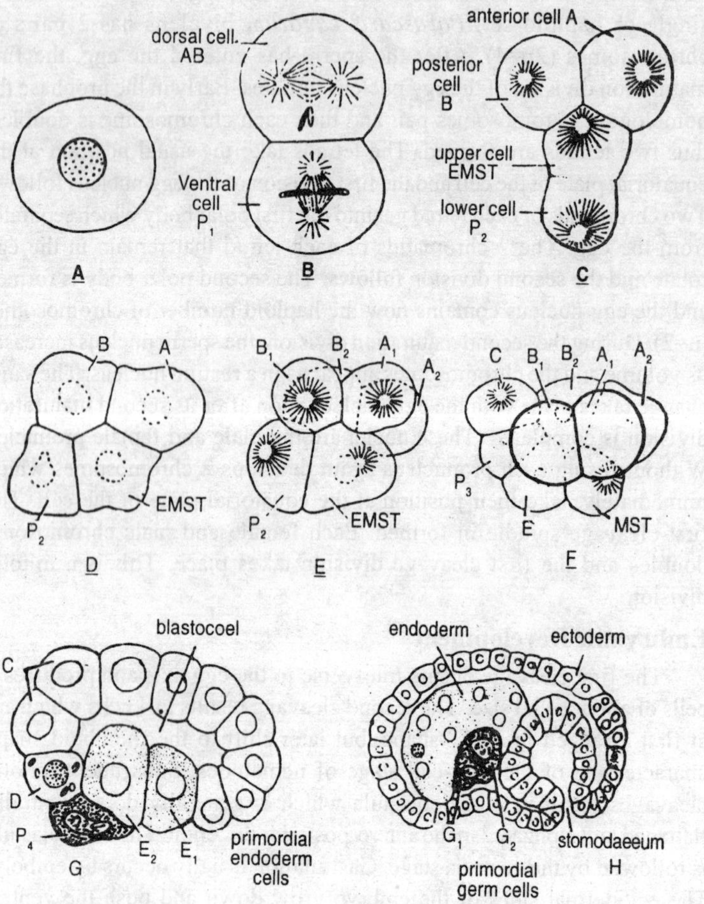

Fig. 13.3. *Ascaris.* Stages showing early development. A—Zygote, B—2-cell stage, C—4-cell stage (T-shaped), D—4-cell stage (rhomboidal), E—6-cell stage, F—8-cell stage, G—Median sagittal section through blastula, H—Median sagittal section through the embryo after invagination of stomodaeum and the primordial germ cells.

development of the embryo. There is little if any evidence of regulatory power. If one blastomere is destroyed in the two-cell stage, the other develops as it would have in a normal embryo. As a result the embryo dies at an early developmental stage. Furthermore, the segregation of the germ cell line from the somatic cells takes place at an early stage in embryonic development. In *Parascaris equorum*, as well as in other nematodes, the germs cell line can be distinguished by the behaviour of its chromosomes. During the early cleavage divisions in the embryo, a peculiar fragmentation of the chromosomes takes place in these nematodes. In the second to the fifth cleavage mitoses in all cells which will form somatic tissue the middle part of each chromosome breaks up into a number of portions. These short segments behave as entirely separate chromosomes in the somatic cells. The remaining club-shaped end portions of the broken chromosomes are left in the cytoplasm where they eventually degenerate. This process of chromatin loss from the nuclei of somatic cells is termed chromatin diminution.

During the late embryonic development cell multiplication ceases in all organs except the reproductive system. The number of cells or nuclei present at that time remains constant throughout the life of the nematode. This condition is known as cell constancy. Subsequent growth of the nematode is accomplished by enlargement of cells. There is also a tendency towards the development of syncytia.

Post-embryonic Development

Having completed their development within the eggs, the juvenile worms hatch. These juvenile worms or larvae are usually fully developed except for body size and gonad development. The larvae grow and undergo some changes in form especially of the anterior and posterior ends and the gonads develop gradually. A remarkable feature of this post-enbryonic development is the occurence of moults. Periods of growth are separated by moults in which the entire cuticle, including the cuticular lining of stoma, oesophagus, vagina, rectum and cloaca is shed. There are probably always 4 moults. The nematode emerges from the fourth molt as a fully developed adult. The adult may continue to grow to some extent.

In some cases, 1 or 2 moults occur within the egg so that the second or third stae larva hatches. The first moult occurs within the egg in most species of the Tylenchida. In other cases the larva undergoes the second moult but does not shed the cuticle. The third stage larva remains in the cuticle of the second stage larva. This cuticle serves as a cyst and such larvae may resist adverse conditions. They are termed ensheathed larvae or "Dauer" larvae.

It is possible to determine the sex of the future nematode during the early stages of the post-embryonic development. In *Heterodera* and *Ditylenchus*, males and females can be distinguished in the third larval stage. In *Tylenchulus semipenetrans* and *Meloidogyne*, females and males can be distinguished in the second larval stage.

PARTHENOGENESIS AND THE ROLE OF THE MALE IN REPRODUCTION

Parthenogenesis

Parthenogenesis is the development without fertilization, i.e. the egg of an animal develops without fertilization by the sperm. Parthenogenetic reproduction occurs in the absence of the sperm and is of two types; *diploid* and *haploid parthenogenesis*. The diploid type is commonest and occurs in some invertebrates including nematodes. Haploid parthenogenesis occurs in some Hymenopteran and Homopteran insects.

The Haploid Parthenogenesis

The haploid parthenogenetic egg is an asexual egg which undergoes complete reduction and capable of fertilization but may develop without any process of syngamy. Such eggs most commonly develop with the haploid number of chromosomes and in this case, as far as known, always produce males which themselves are of haploid constitution.

The Diploid Parthenogenesis

In diploid parthenogenesis the egg typically undergoes no general reduction and hence develops with the diploid number of chromosomes and so far as known is incapable of fertilization and even if same took place the product would be triploid and hence an abnormal zygote. *Paula Hertwig* in her examination of *Rhabditis pellio* found a mutant which produced only females and formed only one polar body without chromosomal reduction and developed with the diploid number of 14 chromosomes. However, development failed unless the sperm entered the egg although it took no part in the subsequent processes. In her opinion this was not a case of fertilization but rather impregnation of the egg by the sperm and therefore the end result was not a triploid.

The general opinion is that diploid parthenogenetic animals were originally bisexual with sexual eggs capable of fertilization. Normally in diploid parthenogenesis there is no reduction in the diploid number of chromosomes and only one polar body is produced. The diploid type of parthenogenesis, even in the sexual egg after completion of reduction, is possible. In a species of starfish the polar nucleus of the second polocyte unites with the egg nucleus subsequent to maturation and the diploid number of chromosomes is present. There seems to be no doubt that the doubling of the haploid number due to fusion of the egg nucleus and the second ploar body occurs in some forms of parthenogenesis. In diploid parthenogenesis the offspring may be either male of female as demonstrated by experiments carried out with gall flies, aphids and daphnid while in haploid parthenogenesis only males are produced.

Without exception, as far as known, female-producing parthenogenetic eggs develop with the diploid number of chromosomes. However, recently workers have uncovered new aspects as to the male producing parthenogenetic eggs of this type. In some invertebrates, males produced through the diploid parthenogenetic process, have one or two fewer chromosomes than the females. Therefore, reduction apparently occurs in the case of one or two chromosome pairs. This occurs in insects such as aphids. A species of *Phyllozera* during maturation of the male-producing parthenogenetic egg shows two chromosomes lagging behind and failing to enter the egg nucleus and are probably cast out in the polar body.

In general, parthenogenesis is followed eventually by sexual forms, or in essence a parthenogenetic series followed by a sexual series, etc.

Parthenogenetic Evolution

Huxley once remarked that comparative evolution is destined to become as important a part or biology as comparative anotomy. Therefore it is advisable to consider briefly the evolution of parthenogenesis because in nematodes this method of reproduction is present in some species of economic importance.

Diploid parthenogenesis consists of two kinds, (a) complete parthenogenesis where a species consists of exclusively females with rare males and every individual arises from an unfertilized egg, and (b) cyclic parthenogenesis or heterogamy where one or more parthenogenetic generations alternate with a bisexual one. Many parasitic worms have the latter type.

Complete parthemogenesis which may be successful for a time will sooner or later become extinct or parhaps return to sexuality. Bisexual and parthenogenetic races have been found within the same species. However, these should be regarded as "sibling species". *White* (1954) claims that a parthenogenetic species in really not equivalent to a bisexual one since it consists of an indefinite number of biotypes, between which the exchange of genes is not possible and will continue to diverge in evolution as one mutation after another establishes itself in the different lines of descent. Thus, parthenogenetic species which have been existence for some time are likely to show a great deal of variability which will not necessarily be correlated with geographical distribution in the same way as variability of a bisexual form.

Since each individual in a parthenogenetic species is capable of reproducing its prolificity is much higher than that of a bisexual form.

Comparison of parthenogenetic animals with related bisexual forms reveals if these are polyploid in constitution. In *Artemia* this is clearly shown. The bisexual race has a somatic number of 42 chromosomes. Meiosis is normal and at metaphase I twanty-one bivalents are present. In the diploid parthenogenetic race there is no reduction in the 21 bivalents and therefore at eleavage I there are 42 chromosomes present. In the tetraploid pathenogenetic race there is no pairing of the chromosomes during oogenesis and therefore the single maturation division takes place with 84 univalents. Triploid parthenogenetic races have been found.

A very unequal sex ratio does not imply parthenogenesis. *Belar* found that in *Rhabditis monohystera* where males form only 6-7 per cent of the population, unmated females do not produce offspring. The egg undergoes a single maturation division and then is stimulated into further development into a female by the penetration of the sperm which eventually degenerates. If an egg undergoes two maturation divisions it becomes capable of being fertilized and after fusion of the male and female pronuclei a male nematode is produced. In some species of this genus there are no males but the femles form a certain quantity of sperm in theri ovaries. Sperm from this source stimulates the egg to begin development and then it degenerates. In an unnamed species referred to a *Rhabditis XX* the eggs only undergo single maturation division, the chromosomes do not pair during the growth stage and at metaphase each univalent splits as in mitosis. The meiosis in the spermatocyte follows a somewhat different course ; two meiotic divisions occur but since pairing seems incomplete and highly irregular the chromosomes passing into these sperms is variable. Since these sperms only function for a short time and a stimulating agents and play no part in the genetical make-up of the nematode this is probably unimportant. In an evolutionary sense the degeneration of the sperm function has rendered a possible degeneration of meiosis and the regularity of pairing and segregation has disappeared.

In *Rhobditis monohystera* which is bisexual although males are in the minority there is only one maturation division in the egg but pairing of the chromosomes is normal and the 20 chromosomes form 10 bivalents. The somatic number is restored by splitting of the chromosomes which takes place during telophase of the single meiotic division. The degeneration in this species has not proceeded as far as that in *Rhabditis* XIV form of Belar's follows the same course as *R. monohystera* but in "XIV" there are no males and the eggs are capable of developing spontaneously without pseudoreduction.

Parthenogenesis in cyst-forming and root-knot nematodes

The possibility of parthenogenesis in *Heterodera* was first fully investigated by *Ellenby* (1957). He concluded, from the results of his experiments with the golden nematode, *H. rostochiensis*, that parthenogenesis does not normally occur in this species. *Fassuliotic* (1957) and Williams working independently arrived at the same conclusions as *Ellenby* with respect to parthenogenesis in the golden nematode. The only instance of parthenogenesis in the genus *Heterodera* was shown by Ellenby's experiments in 1957. Reproduction in the absence of males has also been observed in *H. trifoli.* Parthenogenesis is of the diploid type in which the diploid number of chromosomes is not reduced during meiosis and only one polar body is produced. All the offsprings were females. There appears to be no pairing of homologous chromosomes.

Reproduction of this species in the greenhouse follows a regular pattern of a generation being produced every forty days. Three pure lines of the clover cyst nematode have been propagated each of which originated from a single unfertilized second-stage larva, to the tenth generation. A single male among the thousands of nematodes produced from these pure lines has been reported.

14

Acanthocephala

The Acanthocephala represents a group of worms parasitic in all classes of vertebrates but esecially in fishes and birds. They show similarities both with the Platyhelminthes and the Nematoda, but no agreement has been reached on their phylogenetic position. They resemble Cestoda in lacking an alimentary canal. The chief diagnostic feature of the group is the presence of an invaginable proboscis armed with hooks from which the common name 'spiny-headed' worm is derived. They range in size from 1.5 mm to over 0.5 m. The majority of species are small, the largest being *Macracanthorhynchus hirudinaceus* form pigs. The females are nearly always larger than the males. The intermediate hosts, in the few species where these are known, are arthropods, and paratenic hosts frequently occur. In their general anatomy, life history and habits they show a remarkable degree of uniformity. Useful reviews of the biology, ecology, physiology and biochemistry of the group are those of *Nicholas* (1967, 1973), *Rothstein* and *Nicholas* (1969), *Bird* and *Wallace* (1969), *Pratt* (1969) and *Crompton*, (1970); morphology and taxonomy are dealt with by *Hyman* (1951) and *Yamaguti* (1963).

AVAILABLE SPECIES

In Europe, both larval and adult material is fairly readily available. Useful host lists in fish in Great Britain and Ireland have been compiled by *Chappell* and *Owen* (1969) and *Chubb* (1970). The following are useful labirotory species:

Moniliformis dubius

The adult occurs in the laboratory rat, with the cockroach, *Periplaneta americana* as the intermediate host ; the maturation time in the rat is five to six weeks. Various aspects of the life cycle have been studied by *Moore* (1946) and *Holmes* (1961).

Polymorphus mimtus

A species common wherever wild fowl or domestic duck have access to lakes or streams inhabited by the amphipod *Gammarus sp.* One advantage in working with *P. minutus* is that the larvae (cystacanths) is gammarids take up a bright orange pigment, with the result that infected specimens may be easily recognized ; infections up to 32 per cent have been reported in some areas.

Table 14.1. Species of *Acanthocephala* recorded in freshwater fishes in Great Britain and Ireland.

Species	*Fish host*
Echinorhynchus clavula	eel, perch, roach, grayling, gwyniad pike, dace, rudd, rudd/bream, trout.
E. salmonis	trout eel.
E. truttae	trout, grayling, dace, roach, pike, salmon, rainbow trout, eet.
Neoechinorhynchus rutili	trout, grayling, salmon.
Acanthocephalus anguillae	perch, powan, trout, rudd/bream.
A. lucii	pike, roach, powan, perch, dace, eel, ruffe, trout, bream, rudd, rudd/bream, tench.
Pomphorhynchus laevis	roach, dace, grayling, eel, perch, bream, minnow, pike, rudd, salmon, stone loach, 3-and 10-spined stickleback, tench, trout, rainbow trout, rudd/bream.

Echinorhynchus truttae

The adults of *E. truttae* occur in brown trout and other fishes but the larvae are not so obviously pigmented as *P. minutus* (above). Coinvasion of the gammarid by larvae of both *P. mimutus* and *E. truttae* is, in fact possible. The life cycle of this species has been worked by *Awachie* (1966).

Acanthocephalus ranae

A relatively common species found is the intestine of the toad *Bufo bufo* its intermediate host is the isopod *Asellus* sp. Its mode of attachment to the gut has been studied by *Hammond* (1967a).

Neoechinorhynchus emydis

A useful species available in the United States in North American freshwater turtles; larvae in freshwater snails. An account of this species has been given by *Goodchild* (1950) and *Hopp* (1954).

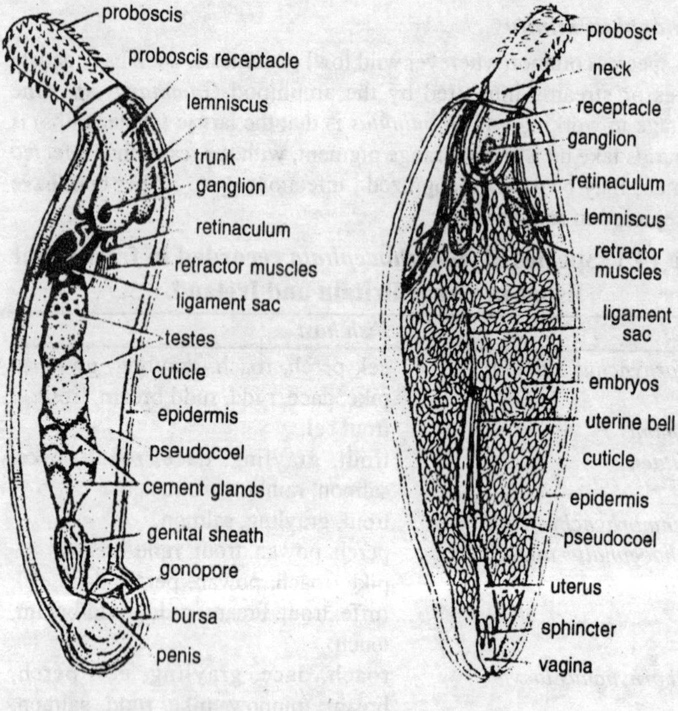

Fig. 14.1. *Acanthocephalus* sp.—morphology of adult male and female.

Macracanthorhynchus hirudinaceus

A large species in wild and domestic pigs with larval stages in scorabaeid beetles. Occasinal infections haven been roported in man.

Pomphorhynchus laevis

A species in dace in the United Kingdom.

Table 14.2. Maturation of Acanthocephala in the definitive host.

Species	*Final Host*	*Prepatent Period*
Leptorhynchoides thecatus	rock bass	8 weeks
Echinorhynchus truttae	trout	9-10 weeks
Polymorphus minutus	duck	22-25 days
Moniliformis dubius	rat	5-6 weeks
Macracanthorhynchus hirudinaceus	pig	10-14 weeks

MORPHOLOGY

Body

The body consists of an anterior *prosoma* with a characteristic spiny *proboscis* and a *neck* free from spines, and a posterior *trunk* or *metasoma*. The spines on the proboscis are symmetrically arranged and of diagnostic value. The proboscis (and sometimes the neck) is usuallly retracted into a proboscis sac by being completely inverted, and the whole prosoma is also retractile. Retraction is carried out by means of retractor muscles. Two lateral protrusions from the body wall at the base of the nack are known as the lemnisci. These are concerned with fluid movement in relation to proboscis activity.

The body wall

The body wall consists of a thick tegument which contains a system of lacunar canals or channels which seem to distrobute absorbed food material. It consists essentially of five layers; the outer most of *epicuticle* appears to contian largely mucopolysaccharide, presumably equivalent to the glycocalyx of cestodes. Beneath the epicuticle lies a tough *cuticle* perforated by numerous pores which in turn lead into canals and ducts which comprise the *striped layer*. The latter merges into a fibrous *felt layer* beneath which is a *radial layer* lacking cell walls but containing nuclei, mitochondria, ribosomes, folded plasma membranes, lipid and glycogen. This layer is probably the most metabolically active part of the body wall. Beneath these layers, a tihck *basement membrane, circular* and *longitudinal muscles* are present.

The above account is based on light microscope studies supplemented by electron microscopy. A number of species have been examined but (as in cestodes) the body wall shows a remarkable degree of uniformity.

Representative ultrastructural sutdies are those on adults : *Polymorphus minutus*, *Moniliformis dubius*, *Acanhocephalus ranae Acanthor*, *M. dubius egg*; *M. dubius*.

Digestive system

This is lacking as in cestodes. Absorption of materials takes place directly through the body wall.

Excretory system

Lacking except in the order Archiachanthocephala in which modified protonephridial organs occur. These consist of flame bulbs (Each consisting of a branching mass of 250-700 flame cells) and ducts which lead into a single duct or bladder.

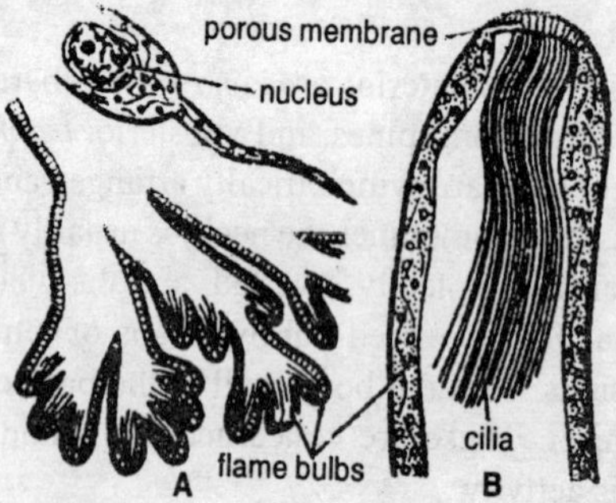

Fig. 14.2. *Macracanthorhynohus*. A—L.S. through the terminal twings of a protonephridium, B—A flame cell in L.S.

Nervous system

This is reduced to a ganglion in the wall of the proboscis sac and two nerve trunks which run posteriorly from the brain obliquely to the body wall. In the male there is a pair of genital ganglia in the penis base.

Nuclear constancy

The group exhibits the interesting phenomenon of nuclear constancy or eutely. In a given species, the number of nuclei in the organs and tissues attained during larval development remain constant except in the gonads. Some nuclei in the epidermis or some species fragment into chromatin material.

Ligament sac and ligament strand

Thie ligament sacs are hollow tubes of connective tissue peculiar to the *Acanthocephala*; these enclose the gonads. In front they are attached to the proboscis sac or the adjacent body wall and posteriorly they connect to some part of the reproductive system. The mumber and arrangement of the ligament sacs varies in the different orders. Thus there are two (a dorsal and a vantral) in the females of the Archi- and Eocanthocephala, But in the males of the same orders only the dorsal sac is present. In the Palaeacanthocephala, there is a single ligament sac in both sexes.

The ligament strand is a nucleated strand to which the gonads in both sexes are attached. This strand is belived to represent the endoderm, and the body space between this and the body wall may be considered a pseudocoel.

REPRODUCTIVE SYSTEM

Male reproductive system

There are two *testes* attached to the ligament strand and enclosed in the ligament sac. In the Archiacanthocephiala and the Palaeacanthocephala, there occur cement glands six to eight in number with ducts leading into the common sperm duct. In the Eoacanthocephala, however, the cement glands form a syncytial mass form which a duct enters a cement reservoir form which in turn ducts enter the sperm duct. The secretion of the cement glands seals the everted male bursa around the female gonopore.

Continuous with the ligament sac is a muscular tube the *genital sheath* which encloses the posterior ends of the sperm ducts, the cement glands and the protonephridial canals when present. There is a short penis which projects into hemispherical or elongated bursa, a cavity made up of the inturned body wall with a proximal muscular thickening the muscular cap. The bursa can be everted to the exterior and is used in holding the female during copulation.

Female reproductive system

This shows a number of peculiarities. The *overy* or *ovaries* (two) which are visibly intact only early in life, break up into ovarian balls each consisting of a syncytium from which ovogonia are formed. These soon come to be free in the palacanthocephals the sac ruptures and they fill the body cavity. The ligament strands lead to a muscular organ termed the *uterine bell*. Posteriorly this opens into a uterine tube bearing two bell pouches, which connects to a muscular uterus and a non-muscular vagina opening to the exterior. The vaginal aperture is guarded by a sphincter.

The uterine bell serves as a selective apparatus. Ovic larvae are taken in by the uterine bell and passed through the uterus to the vagina and so to the exteror. Immatue eggs are rejected and returned via and aperture in the ventral bell to the dorsal pseudocoel. The selective mechanism in *P. minutus* appears to operate on egg length, most eggs less than 100 μm being immature. The uterine bell in continuous with the dorsal ligament sac in the Archi- and Eoacanthocephala. In the Palaeacanthocephala the single dorsal strand connects to the inner side of the uterine bell. In *M. dubius* a XO-sex determination mechanism appers to operate with 2n=8 (female); 7 (male).

Copulation and fertilization

The male bursa is used for grasping the female and the penis is inserted into the vagina and sperm discharged. The secretion from the cement glands is then poured over the whole posterior tip of the female and the gonopore securely sealed. The secretion persists for a time as a hard brown 'copulation cap'.

Structure of eggshell

Four egg membranes have been desribed in most species three in others. In *M. dubius* the outermost layers contain keratin, the third is PAS positive and made up of proteinchitin. The innermost layer is proteinaceous. The eggs of some species are extremely resistant, eggs of *P. minutus* retaining their infectivity for 2.5 months at —22°C.

LIFE CYCLE

The life cycles are known in detail for only a few species. the intermediate hosts are insects (larvae, beetles, cockroaches) for those parasitic in land animals and crustaceans or molluscs for those in aquatic animals. Some developmental stages of *Polymorphus minutus* are shown. The time for hatching of the eggs varies with species, two hours being the earliest reported time for *M.dubius* in the cockroach eggs of *Neoechinorhynchus saginatus* hatch in ostracods within one hour.

Hatching is markedly affected by pH, a pH > 7.5 being necessary for optimal results; a molarity > 0.2 is also essential. In the presence of carbon dioxide, however, hatching takes place at a lower pH. A chitinase appears to be involved in the hatching process and bile salts appear to be the chief agent 'activating' larvae in the definitive host.

CLASSIFICATION

Three orders are recognized :

Order 1 Archiacanthocephala. Hooks on proboscis either in long rows or in a few circles ; spines lacking on trunk; main lacunar vessels dorsal and ventral both dorsal and ventral ligament sacs present in female cement glands separte, with eight present in males; in terrestrial hostes e.g., *Moniliformis dubius*.

Order 2 Palaeacanthorcephala. Hooks on probosicis usually in long rows; spines on trunk ; nuclei in hypodermis usually fragmented; main lacunar vessels lateral; ligament sac in female single; cement glands separate; mostly aquatic hosts e.g. *Leptorhyncohoides thecatus* (in fish; larvae in amphipods).

Order 3 Eoacanthocephala. Hooks on proboscis usually in a few small circles ; few nuclei (large) in hypodermis; main lacunar vassels dorsals and ventral; dorsal and ventral ligament sacs in female; cement glands syncytial; in aquatic hosts, e.g., *Neoechinorhynchus emydis*.

Several alternative classifications to the above are also in use.

Revision Questions

1. Describe the structure and classification of Acanthocephala.
2. Draw neat and labelled diagrams of male and female *Acantocephalus* (Description is not required).

15

Parasitism and Parasitic Adaptations

Parasitism is such an association between two animals of the same or different species in which one lives on or inside (parasite) the body of the other (host) receiving nourishment and shelter without any compensation for the host. Therefore, in this association the parasite is benefited and the host is harmed.

ORIGIN AND EVOLUTION OF PARASITISM

Parasitism is a secondary mode of life and has arisen from pre-living condition. Parasites living inside the bodies of their hosts (*endoparasites*) are the most successful ones. Between the endoparasite life and free-living life, intermediate stages are not uncommon, proving that parasitism did not evolve all of a sudden but followed a series of steps in succession.

Living together or *symbiosis* started when smaller animals occasionally taking shelter temporarily on the body of some larger form. When this temporary shelters was repeatedly used, the association of two organisms become more intimate and the small oraganism gets not only the shelter but also free transport and nourishment in the form of bits from the host, without interfering with the activities of the host or injuring it. Such a simple symbiosis is known as *commensalism* and the benefactor is called a *commensal*.

Still later, the commensal started living within the body of the host without doing any harm to the host. Thus a commensal may be firstly an *ectocommensal* and later an *endocommensal*. Symbiosis sometimes proves beneficial to the host. Thus both the organisms benefitting each other mutually, and the host becomes so dependent upon the commensal that it cannot get along without it. This kind of association is called *mutualism*.

Commensalism usually evolves not in the direction of mutualism but towards a deeper association, the *parasitism*. A commensal that at first takes shelter, then bits of food and finaly starts to feed on body tissues of the host so that the latter has to suffer with some harm.

Different types of parasitism

The parasitism can be classified into following categories :

1. *Accidental*. When a free living animal accidently reaches inside the body of the host then it leads a parasitic life for some time.
2. *Facultative*. When the animal can live both as parasitic or free-living, they are known as facultative parasite.
3. *Obligatory*. These animals can not lead a free life, therefore, they need a suitable host for them. The obligatory parasitism is of different types :
   (a) *Ectoparasitism*. The animal lives on the external surface of the body of the host.
   (b) *Endoparasitism*. The animal lives inside the body of the host in various organs and systems.

Effect of parasitism upon the parasite

Changes from a free-living condition to a parasitic life involves changes in the body of the animal. Due to parasitic mode of life certain new structures are developed due to new environmental requirements similarly certain characters, which are of no use in new environment, are lost, All such changes are included with effect on parasitism on the parasite. The helminths are modified morphologically as well as physiologically to live in their particular environment.

I. Morphological Adaptations
II. Physiological Adaptations

I. MORPHOLOGICAL ADAPTATIONS

These are of two types (i) Structure which are partially or wholly lost. (ii) Structures which have developed a fresh.

Degenerations

1. ***Organs of locomotion.*** Locomotion is actually an effort for procuring food. But parasites habitually inhabit such places in the host's body, where sufficient food is available without effort. Thus, the organs of locomotion, such as the cilia of free-living turbellarians, are absent in the parasitic forms. It is interesting to note that the locomotory organs are duly present in free-living larvae of parasitic forms; the miracidium possesses cilia and the cercaria possesses a tail for locomotion.
2. ***Organs of nutrition***. Food of the parasite comprises the readily available digested or semidigested food of the host. Therefore, there is a reduction in trophic organs. In trematodes, there is an incomplete

gut and in most of them, suctorial pharynx is present. In cestodes, the mouth and alimentary canal are lacking because digested food is readily available in the gut of the host.

3. ***Nervous System.*** Living safely in a more or less stable environments, parasites do not need a complex nervous system.
4. ***Sense Organs.*** Progressive parasitism is accomplished by the loss of sense organs. The sense organs, necessary for quick and effcient response to the stimuli, are associated with the free active life. In the host the environment is more or less uniform and so the sense organs are not essential. Hence they are reduced. The endoparasites are generally provided with tangoreceptors.

Neoformation

1. ***Protective covering.*** There are no epidermal cells in the adult, instead of which the body is covered by a several-layered thick cuticle which protects the parasite from the juices of the host.

   The outer surface of the cuticle is formed of a fibrous protein called *keratin*. The cuticle may be provided with scales.
2. ***Organs of attachment.*** For a firm grip on or in the host's body, some special organs of adhesion are necessary. The flatworms, for this purpose, are variously armed with suckers, hooks and spines. The suckers themselves may be wi^th or without hooks and spines.
3. ***Organs for penetration.*** The parasites must penetrate the host's body for entering it. So the parasites develop certain structures for penetration. Miracidium larva has a conical process at the anterior end called *apical papilla*. There are a pair of penetration glands inside the body near the anterior end.
4. ***Suctorical pharynx.*** All the endoparasites except cestodes have a suctorial pharynx by means of which liquid food is sucked into the intestine. Intestine is always branched.
5. ***Reproductive system.*** It is higly developed and the production of eggs is prolific. The parasitic flatworms are hermaphrodite. In cestodes the reproductive organs are much more elaborate and are repated in each proglottid. Each mature proglottid possesses one set (*Taenia solium*) or two sets (*Diphylidium*) of male and female reproductive organs. In each gravid proglottid all other oragns of the system degenarate to make room for the uterus which is greatly enlarged and branched to accomodate a large number of eggs.

II. PHYSIOLOGICAL ADAPTATIONS

1. ***Protective mechanism.*** In the alimentary canal of the host the parasite protects itself from the action of digestive juices of the host.

The tapeworms accomplish this (a) by stimulating the walls of the gut to secrete mucous, which then forms a protective clothing around the parasite, (b) by secreting antienzymes to neutralize the digestive enzymes of the host, and (c) by probably continually renewing their protective body covering *i.e.* tegument. Lime cells in the body wall of tapeworms nutralize the acids produces by the host. Blood parasites are known to withstand the effect of antibodies and phagocytes of blood by some mechanism.

2. ***Osmoregulation.*** The osmotic pressure of body-fluid of endoparasites, especially in case of trematodes, is almost the same as that of the host so that there is no necessity of osmoregulation. But in the intestinal tapeworms, the osmotic pressure is slightly higher than that of surrounding medium. This permits absorption of nourishment from their general body surface from the surrounding medium.
3. ***Anaerobiosis.*** The *pH* tolerance of the parasite is high from 4-11. The intestinal parasite lives in an environment completely devoid of oxygen. The respiration is of the anaerobic type consisting of extracting oxygen from the food stuffs. In the absence of oxygen energy is obtained by the fermentation of glycogen which is broken into lactic acid.
4. ***Periodic appearence.*** There are some parasites which appear at definite period. The larvae of *Wuchereria bancrofti* circulate into the peripheral blood circulation in the night between 10 P.M. to 4 A.M. for further development they require *Culex* mosquitoes. The *Culex* bites a man at night *i.e.* nocturnal in habit. Their migration is correlated with the nocturnal habit of *Culex*.
5. ***Neoteny.*** It is a special phenomenon in which the larvae are capable of reproduction and they produce young ones. *Ligula* (Cestoda) exhibits neoteny and this can be considered an adaptation for parasitic mode of life.
6. ***High fertility.*** Parasites develop enormous fertility producing millions of fertile eggs. This is correlated with the passive transference of the infective stages of the parasites from one host into another. While passing through the complex life-cycle, these potential offsprings face several hazards as a result of which a very small percentage of the total eggs produced reaches adulthood. For example in *Fasciola*, if the eggs fall in water, the first larva (miracidium) finds a suitable snail within a limited period, the cercaria encysts, on vegetation within the reach of final host and if the metacercaria happens to be ingested by the proper final host, only then the life-cycle is completed. Not

only this, the larval stages themselves multiply say for example for sporocyst produces a number of rediae. A single redia produces a number of cercaria larvae. Similarly a single hexacanth may multiply by generating buds which produce daughter and grand daughter scolices (Hydratids of *Echinococcus*).

7. ***Transferance of eggs or infective stages.*** This is another major problem for which the parasites have to develop adaptations. Transference from one host to another is either :
   (i) **Active transfer.** This is a less common method. The infective stages are actively transferred to final host. The larval stages of most parasites bore directly through the skin of their host such as the larvae of hookworms and schistosomes. OR
   (ii) **Passive transfer.** This is the most common method and usually results from the careless habits of the host and is achived in several ways such as contamination, contagion and inoculation. Oral infection occurs when the host takes uncooked food or contaminated water. The intermediate host, such as mosquitoes, inoculate parasites when they suck the blood of hosts.

Effect of parasitism on the host

Although a given host may immune to a parasite by virtue of being constitutionally unstable as a habitat, it is important to recognize that the establishment of a parasite in a host also involve the capacity of the parasite to resist reactions of the host that might be deleterious. These host reactions would include inflammatory tissue response and immune responses of a humoral character. Both types of host reactions occur in most parasitisms involving characters. The immune system in higer vertebrates is composed of several tissue components. These include fixed cells of spleen, liver, lymph nodes and thymus. There is also a system of mobile phagocytic cells, originating mainly in the bone marrow including the heterophil and eosinophil leucocytes. In addition, there are a number of potential macrophage cells in the body, such as undifferentiated mesenchyme cells. Some of these cells has the ability to elaborate antibodies, which may be released from cells into the body fluid. These antibodies may combine with parasitic organisms or their products and produce an inhibitory effect. A peculiar type of immunity develops in man due to the presence of tapeworms; after the first infection no further infection is possible as long as the primary infection lasts, this is termed *premunition*.

The possible role of interferons in the immunity of hosts of animal parasites should be also mentioned. These substances are produced in rcsponse to the presence of a foreign agent. Unlike antibodies, they seem to render host cells until as habitations for intracellular parasites.

There are other types of host response that cannot be clearly recognized as immune in characters. For example, the changes in the fatty acids found in the alimentary canal of animals infected with the tapeworm, *Hymenolepis,* or the marked modification of the amino acid composition of bile in sheeps harboring *Fasciola hepatica* are obviously host response to the presence of parasites concerned. It would be of great interest to determine whether such responses are beneficial to the parasites, to the host or to neither.

Revision Questions

1. Give a general account of parasitism and parasitic adaptations in Helminthes.
2. Describe parasitic adeptations in *Fasciola* and *Taenia.*

Index

E

F

G

H

J

K

L

M